ori" (Humming Bird)

similarity to the C 30 autogiro being built under license by Focke-Wulf at that time. Power was provided by an Sh 14 engine driving a conventional propeller. (5, 5a)

Fl 185:

The development contract for the Fl 185 was issued by the RLM in February 1937; two SV-Types were planned. It was given the designation "Helicopter Conversion 184" in the Aircraft Development Program, because the planned Fl 184 V3 was used in the construction of the Fl 185 V1. The project was financed with the insurance sum received from the destroyed Fl 184 V1 and a grant of approximately RM 50,000 from the RLM. Mockup inspection took place in April 1937.

The Fl 185 was a gyroplane equipped with a three-blade rotor; torque balance was achieved by means of variable-pitch pusher propellers driven by extension shafts located at the ends of outriggers mounted on each side

Anton Flettner in front of the Fl 282 V14 CJ+SH in Schweidnitz, October 1944.

Flettner's first experimental helicopter during tethered flight (1934). Note the bracing tower above the cockpit.

Flettner autogiro, Fl 184 V1 D-EDVE.

of the fuselage. The starboard propeller produced backward thrust, the port forward, so that the total moment exerted on the fuselage balanced that of the main rotor.

The sole prototype built was the Fl 184 V1 D-ELFT, which carried out a number of flights in 1938. Results were favorable, although all tests were made in close proximity to the ground.

Fl 265:

The future concept of the Fl 282 was apparent in the Fl 265, which was conceived and designed in 1938. The Fl 282 exhibited significant improvements compared to preceding designs, for example the double rotor with intermeshing blades, automatic changeover from helicopter to autogiro flight (autorotation, from the V4), and the ability to switch back to helicopter flight. Six prototypes were built (Werk-Nr. 1579 - 1584), one of which (TK+AN) was tested in the large wind tunnel at Chalais Meudon (paris) in the summer of 1940.

The Fl 265 V1 D-EFLV took off on its maiden flight in May 1939 in the hands of Flugkapitän Richard Perlia following an extensive series of ground tests (first full-power run on 17/1/39, first tethered flight at a height of 20 cm. on 14/3/39). It was also used for extensive military testing in conjunction with the Kriegsmarine (for example deck landings on an approximately 25-meter-square landing platform on the cruiser Köln). The aircraft was removed from service on April 29, 1940.

The Fl 265 V3 crashed due to pilot error on August 21, 1939 and was destroyed in the ensuing fire. The Fl 265 V5 (retired on 26/11/41) and the Fl 265 V6 (retired 25/3/42) were still on hand at the Flettner Company in November 1942. Their return to service was postponed and eventually abandoned. (7)

All the Fl 265 prototypes accumulated the following in the course of testing: pure flight time 126 hours, 32 minutes; takeoffs 1,180; running time 518 hours, 44 minutes (flight time + tethered flights + engine run-up time). At the time this represented a record number of hours of operation for a helicopter.

Fl 282:

The Fl 282 was undoubtedly Anton Flettner's most successful helicopter development. Today one can say without exaggeration that it was the best helicopter of its day.^902

The exact date on which the RLM issued the development contract for the Fl 282 is not known. It can be assumed with certainty, however, that design work was begun in 1939 while testing of the Fl 265 was under way. All the test results and experience from the Fl 265 were applied to the design of the Fl 282.

By mid-August 1941 work on the Fl 282 V1 had progressed to the point where it could be used on the ground as a transmission unit test bed. The test bed was anchored but could climb the length of the restraint to a certain height and hover. 125 hours and 39 minutes were spent in these tests, which lasted until November 21, 1941. In the event of the destruction of the V1 or a lengthy interruption of testing due to some other problem, it was planned to use the Fl 282 V4 as a replacement test bed.

Fl 282 V2 with fully glazed cockpit. First flight October 30, 1941.

Fl 265 TK+AN in the Chalais-Meudon wind tunnel (Paris, 1940).

Fl 265 ??+SA during trials on the cruiser Köln.

The first free flight was made by the Fl 282 V2 on October 30, 1941. At the controls was test pilot Ludwig Hoffmann, who had joined Flettner as the successor to company pilot Perlia. Tests resumed in March 1942 following design changes to the cardan shaft; the tests revealed a significant improvement in the helicopter's handling. The Fl 282 V2 was taken out of service on May 25, 1942; the aircraft's transmission and engine were removed for use in other prototypes.

Hoffmann carried out two altitude flights in the Fl 282 V3 on April 27, 1942. On the first he reached a height of 3,500 meters over the takeoff point and on the second (from 1517 to 1610 hours) 3,800 meters in 36 minutes.

The Fl 282 V5, which featured a number of changes compared to the V3 (tail surfaces, fuselage, pilot seat), began flying in January 1942.

It also remains to be mentioned that the Fl 282 V2 and V3 had a fully-glazed cockpit. From the V5 on the pilot seat was either entirely open or had a plexiglass shield in front or on the sides.

Various empennage arrangements were also tested. The V3 had a horizontal stabilizer right and left with end fins. A V-shaped stabilizer, similar to that of the Cierva C 30, was installed on the V23.

Calculations revealed that reversing the direction of rotation of the rotor should significantly improve directional stability, especially at low power settings. The V8 was rigged up as a test bed to confirm this. Flight tests with the V9 and V15 began in September 1942 and confirmed the theoretical findings. Directional stability was so good, that the entire fuselage section aft of the firewall was simply left off the V9. Flight tests were problem free. The V9 was the prototype for a version to be operated by the Navy from large submarines and designated "Modified Model in Form of a Standing Cylinder" (Fl 282 U).

The Fl 282 V21 and V23 were built as two-seaters; an observer sat with his back to the pilot in the fuselage behind the engine block. This made necessary the removal of the fuselage fuel tank; it was replaced by two unprotected, cylindrical tanks mounted externally on both sides of the pilot seat. Furthermore, a different rotor arrangement – two three-blade rotors – was tried out on the test bed. It proved to run extraordinarily smoothly, however this was not a consideration for military use.

The entire Fl 282 flight test program was not conducted at Rechlin, as was customary for land-based aircraft; instead, from August 1942, trials were carried out at E-Stelle See Travemünde. In charge of helicopter and autogiro testing there was Fliegerstabsingenieur Dipl.Ing. Hans Fischer; his assistant was Fliegerstabsingenieur Dipl.Ing. Otto Dumke. After Fischer was badly hurt in the crash of one of the E-Stelle's Do 217s (RB+YH), Fliegerstabsingenieur Dipl.Ing. Gerhard Geike assumed responsibility for the workings of this group.

Fl 282 V2 with fully glazed cockpit. First flight October 30, 1941.

Fl 282 V3, GF+YC, horizontal stabilizers with end plates, glazing in front of pilot only.

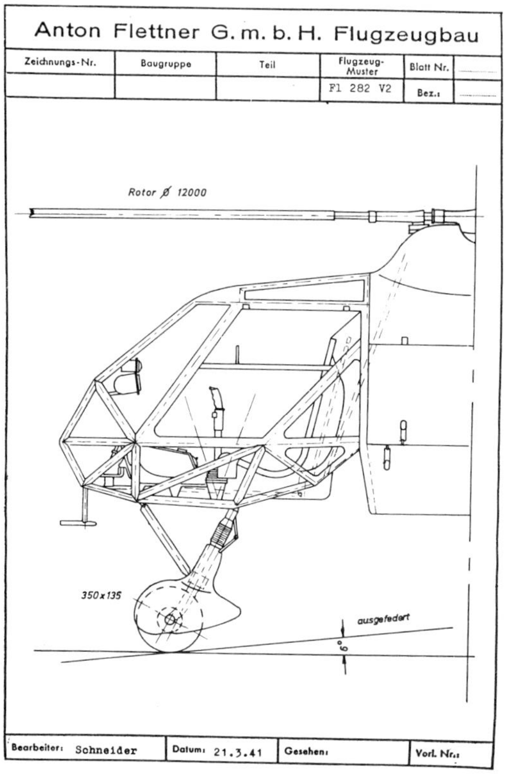

Drawing of the cockpit glazing of the Fl 282 V2.

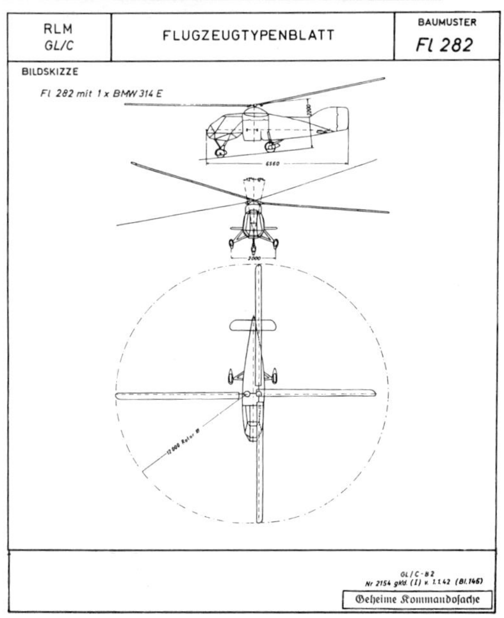

Official RLM type sheet (Sheet No. 145), showing the Fl 282 V2.

Fl 282 V21, CI+TU, a two-seater, with company pilot Fuisting at the controls. In the rear seat is Anton Flettner.

All photos from the Mohr Archive, background photo Pages 2 and 3 from the Birkholz Collection.

Fl 282 V21, CI+TU, a two-seater. Note the cylindrical fuel tank next to the cockpit. The 017 on the rotor head is the transmission number.

Fl 282 V6, GF+YF, during trials with "towed stores", May 1943.

Travemünde was selected because the air traffic safety ship Greif was based there and could be used for deck landing trials. The site also simplified helicopter sea trials with the navy. It was also imperative to move the site of the trials from Flettner's facility in Berlin, where there was a greater risk from the growing Allied bombing raids.

One special experiment took place in co-operation with the DVL's Institute for Marine Aviation. It involved "towing tests with a 50-kg. gliding body." The proposal originated from the director of the Erprobungs- und Lehrkommando (Testing and Instruction Detachment) 20, Hauptmann von Winterfeldt, who carried out the towing flights from an anti-submarine vessel in Gotenhafen on May 3 and 4, 1943. In addition to its role as a reconnaissance aircraft in support of the sub-chasers, armed with bombs the Fl 282 was also to participate actively in the anti-submarine role.

The following extract from the E-Stelle Travemünde's monthly report of June 1944 is also worthy of mention:

"A mock combat between the Fl 282 (Flettner company pilot Fuisting) and a Fw 190 (pilot Ltn. Eisenlohr of E.Kdo.25) took place at Schweidnitz on 22/6/44, in order to investigate the chances of a fighter hitting a helicopter. At present the evaluation of the film and the pilot reports have not yet arrived. At heights above 100 meters the fighter was able to get the helicopter in its sights briefly. Near the ground, especially in difficult terrain, the fighter has little chance against a helicopter."

Vulnerability to gunfire was also investigated, whereby they proceeded on the assumption that the mathematical probability of a moving rotor blade being hit was much less than that of a fixed wing. Another consideration was that it should be extremely difficult to fire on and hit the slow-moving helicopter from a fast fighter. The latter could escape by making brief evasive movements which the fighter could not follow. Furthermore, tests involving ground firing at the moving rotor blades were carried out, as the helicopter was felt to be more vulnerable to gunfire from the ground than from the air. An unmanned, tethered Kolibri was used; in spite of several hits in the rotor blades ground fire failed to bring down the helicopter.

The "Anton Flettner G.m.b.H." was a small engineering company dedicated to helicopters. It is believed that the firm was founded in Berlin in 1935. The earliest document the author has been able to find is a letter from the Military Economics Inspectorate (W.I.) III, Berlin to the RLM concerning firms involved in production for the Luftwaffe and dated October 2, 1936. The letter states that the W.I. III first became aware that the firm had been given

Fl 282 V12, CJ+SF; taxying the helicopter under its own power was strictly forbidden.

The Kolibri Described

important work by the RLM (LC II) as the result of a formal application for an exemption from military service for one of Flettner's skilled workers. (9)

The same inspectorate sent the secrecy agreements to the Flettner Company, Berlin-Johannisthal, Segelfliegerdamm 27, for signature on January 22, 1937.

Due to the growing number of air raids on Berlin, in August 1943 the company began transferring its operations to Schweidnitz in Silesia (approx. 50 km. SW of Breslau); due to the deterioration of the transportation system the operation took several months. The Fl 282s on hand with the company were also flown to Schweidnitz to continue the test program. In February 1944 the work force reached approximately 120 men, its highest level ever.

With the Red Army approaching Silesia, in January/February 1945 the company moved back to Berlin-Tempelhof. Any systematic work or further production was of course out of the question under these circumstances. To make matters worse, two days after its arrival the rest of the company's equipment was destroyed in a night raid on Tempelhof. What was left of Flettner was subsequently evacuated to Bad Tölz (Upper Bavaria); two Fl 282s were also flown there. The history of the Anton Flettner G.m.b.H. as it pertains to this documentation ended there with the arrival of American troops.

Technical description of the Fl 282 B-0 and B-1 (as of 23/12/1943

1. Purpose: I. Reconnaissance aircraft (land), II. Shipboard reconnaissance aircraft.

2. Designation: Fl 282 B-0 (with no cockpit glazing), Fl 282 B-1 (plexiglass glazed canopy).

3. Crew: 1 pilot.

4. Design: Single-engined helicopter and gyroplane of mixed construction with two intermeshing, counter-rotating two-blade rotors.

(a) Fuselage: Welded steel tube, truss-type construction. Forward section with open (B-0) or enclosed (plexiglass) cockpit (B-1). Recess for back-type parachute in rear wall of seat. The fuselage center-section contains the entire power plant, the outer skin consists of removable doors, hoods and panels. Oval cross-section. The rear fuselage is fabric-covered with a rectangular cross-section.

(b) Rotor Blades: A tubular steel spar with riveted-on wooden ribs and plywood skin with fabric covering. Rectangular outline with rounded tips. Attached to rotor head with delta and alpha hinges. Receptacle for balance weight at end of spar (scrap). The axes of both rotors are angled outboard at 12 degrees from the vertical. Seen from above, the right rotor rotates clockwise, the left counterclockwise.

(c) Control Surfaces: No elevator. Two-part horizontal stabilizer with single spar. Tubular steel spar with riveted-on wooden ribs. Plywood leading edge, fabric-covering, cantilever construction. Angle of incidence adjustable between -15 and +5 degrees from cockpit. Rudder: wooden construction, fabric-covered, deflection to 40 degrees. Vertical stabilizer: wooden construction with plywood leading edge and fabric covering, bolted to fuselage frame. No ailerons.

(d) Undercarriage: Braced fixed nosewheel with VDM oil shock-absorber leg, coupled with rudder foot pedals (steerable nosewheel). Nosewheel: EC tire 350 X 150 mm, mainwheels: EC tires 465 X 165 mm.

5. Power Plant:

(a) Engine Type: 1 BMW 314 E. Output: 160 H.P., compression ratio 1:6.0, rotor reduction ratio: 1:12.

E-Stelle Travemünde	Vorläufiges Kennblatt Fl 282 B-0 und B-1.	Lfd.Nr. 8 Auftr.E2-75/1 Blatt 5

12. Zusammenstellung der Lasten:

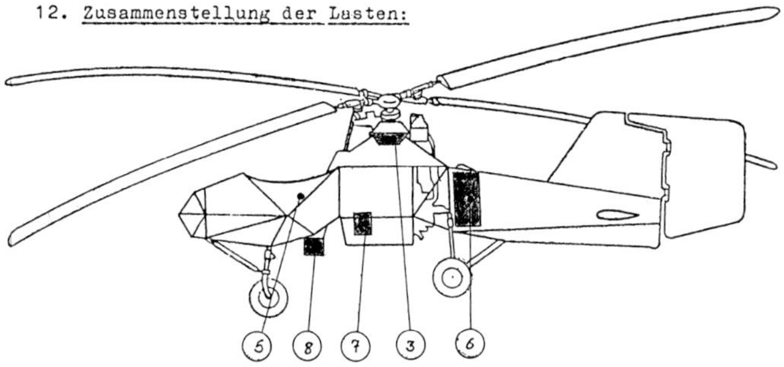

	Baureihe	B-0		B-1	
Verw.Zw.		I kg	II kg	I kg	II kg
1.	Leergewicht	715	715	735	735
2.	Zus. Ausrüstung	61	71	61	71
3.	Schmierstoff im oberen Getriebe 5 ltr.	5	5	5	5
4.	Rüstgewicht	781	791	801	811
5.	Führer mit Fallschirm (und Schlauchboot)	90	95	90	95
6.	Kraftstoff 105 ltr.	75	75	75	75
7.	Schmierstoff 10 ltr.	9	9	9	9
8.	Bomben 2 x 5 kg		10		10
9.	Fluggewicht	955,0	980,0	975,0	1000,0

Gruppe E2d	Bearbeiter Dümke	Tag 23.12.43/Fr.		Rg.Nr.139

Fl 282 weight distribution chart.

E-Stelle Travemünde	Vorläufiges Kennblatt Fl 282 B-0 und B-1.	Lfd.Nr. 8 Auftrag E2-75/1 Blatt 6

16. Schattenrisse.

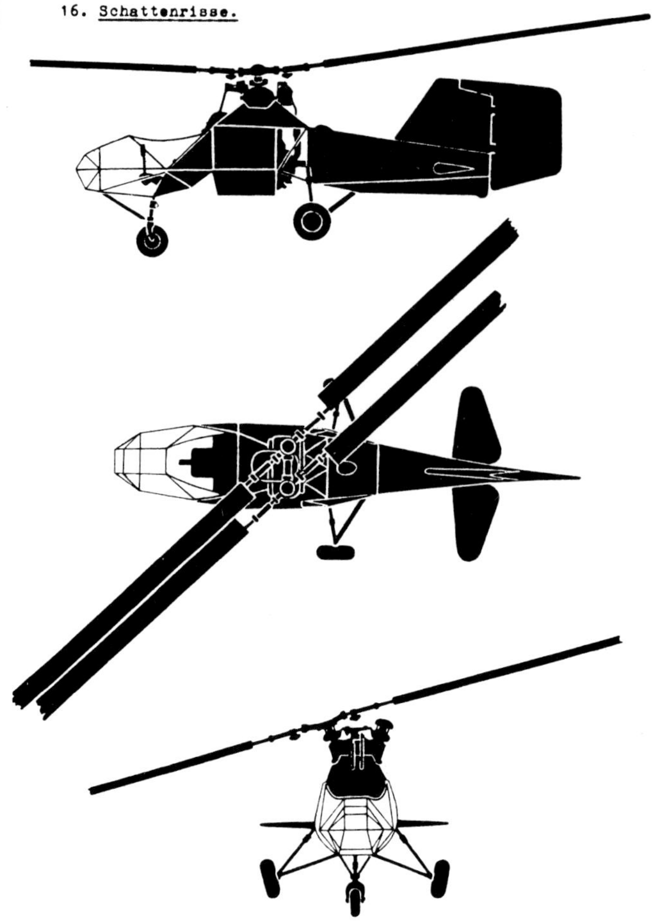

Gruppe E2d	Bearbeiter Dümke	Tag 23.12.43/Fr.		Rg.Nr.139

Provisional identification sheet for the Fl 282 B-0 and B-1 with silhouette drawings.

(b) Transmission: Lower transmission flange-mounted on front of engine, in front cooling fan on extended crankshaft. Upper transmission, linked to the lower by a double cardan shaft. Firewall over and behind the engine.

(c) Fuel Tank: An unprotected metal tank located behind the firewall in the aft fuselage, consumable content 105 liters.

(d) Oil Tank: An unprotected 10-liter metal tank flange-mounted on the lower gearbox. Oil content of upper gearbox = 5 l.

(e) Type of Fuel: 87 octane aviation gasoline.

(f) Cooling System: Continuous cooling provided by an eight-blade wooden cooling fan with direct drive from the engine. Oil cooler which is switched off for cold-weather operation.

(g) Control System: Control about all three axes by means of periodic or constant changes in angle of incidence of the rotor blades by means of stick and rudder pedals, to which the rudder is connected. Change-over from helicopter to autogiro (autorotation) flight and reverse achieved by means of a speed-sensi-

Anton Flettner G.m.b.H. Flugzeugbau

Zeichnungs-Nr. | Baugruppe | Teil | Flugzeug-Muster Fl 282 V 6 | Blatt Nr. | Bez.:

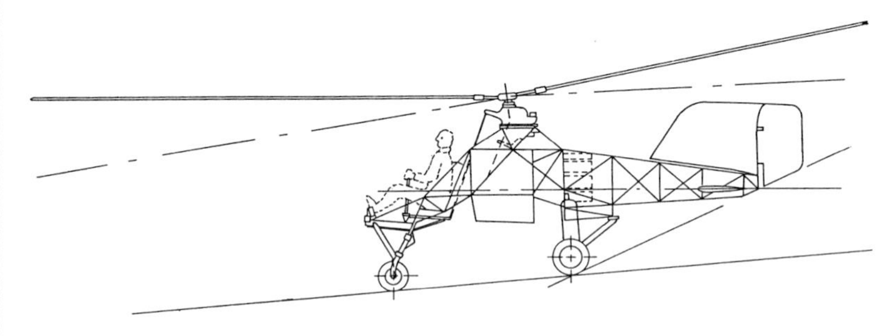

Vorl. Nr.: | Gesehen: | Datum: 24.XI.41 | Bearbeiter: Rehbach

Sketch of the fuselage framework structure.

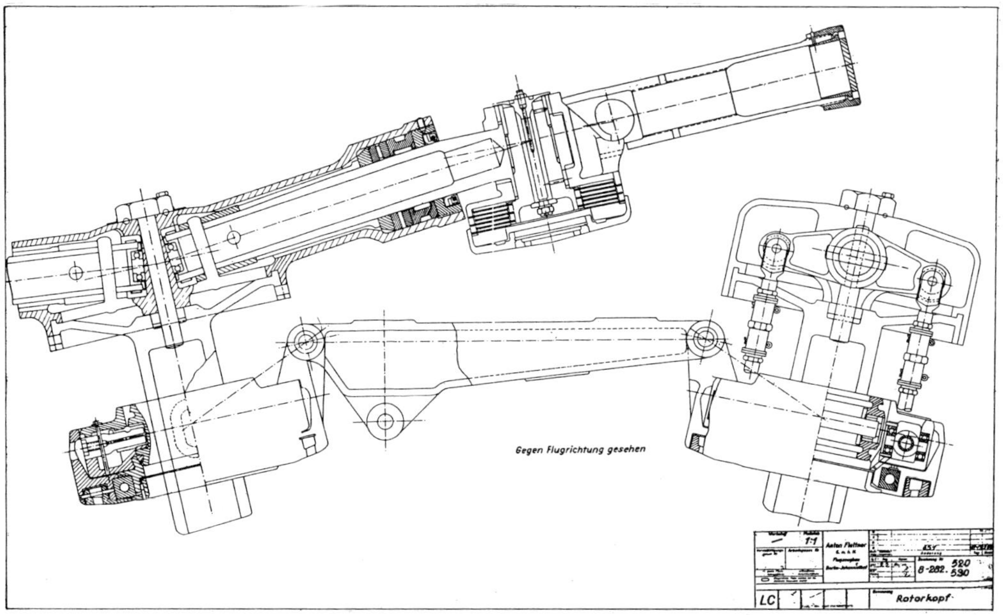

Rotor head, drawing number 8-282 520/530 from February 9, 1941.

tive hydraulic regulator controlled by a hand-operated control lever; change-over to autorotation mode is automatic in the event of engine failure.

7. Equipment:
(a) Flight Instruments and Navigation Equipment: 1 airspeed indicator, 1 altimeter, 1 vertical speed indicator, 1 turn-and-bank indicator, 1 RPM indicator for rotors, 1 rotor blade angle indicator, 1 master compass, 1 dashboard clock.

(b) Safety and Rescue Equipment: 1 back-type parachute, 1 first-aid kit, 1 one-man inflatable raft (shipboard reconnaissance version).

(c) Radio and Signalling Equipment: 1 FuG 19 radio installation, 1 rigidly-mounted single-barrel flare pistol,
1 rigidly-mounted signal lamp (shipboard reconnaissance version).

(d) Bomb-Release Mechanism: 1 bomb magazine for two 5-kg explosive devices (shipboard reconnaissance version), 1 magazine for smoke buoys (shipboard reconnaissance version).

8. Special Equipment: Deck-landing equipment, consisting of: 1 tether with landing cable and electrical release.

9. Dimensions:
Length of aircraft: 6150 mm
Width, including undercarriage and tail surfaces: 2400 mm
Height with rotor blades removed: 2400 mm
Rotor diameter: 12000 mm

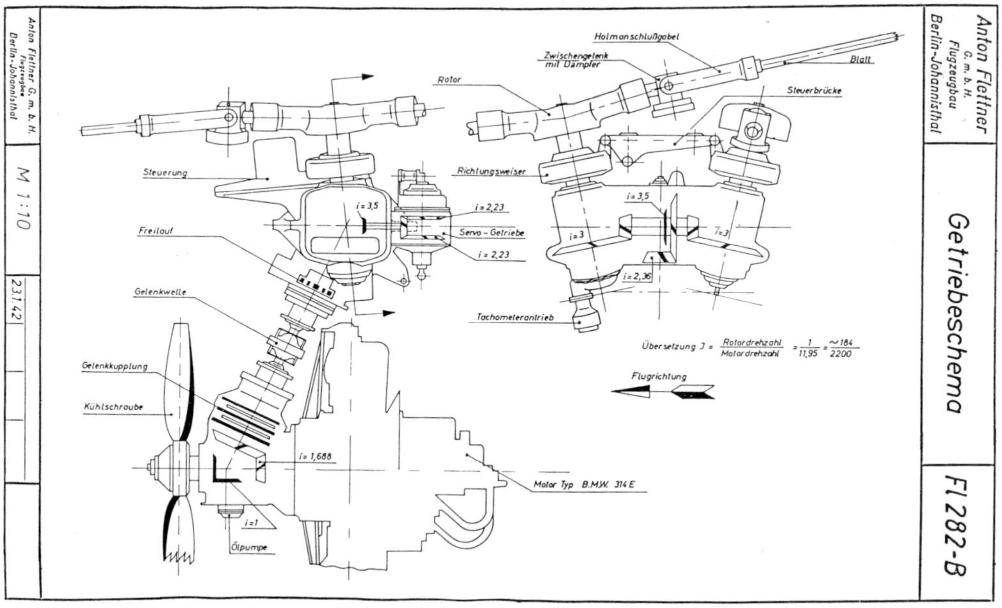

Sectional drawing of the rotor transmission; below left the cooling fan.

Aircraft Program (Prototypes, Anton Flettner) Delivery Plan No. 222 Issue 2, Secret Command Matter 4863/42, from November 1, 1942.

Type	Engine	Prototype	Location	Werknummer	Remarks
Fl 282 B	1 BMW 314 E	5	Flettner	282000005	Flight tests at the E-Stelle Travemünde
Fl 282 B	1 BMW 314 E	6	Luftwaffe	0006	Service trials
Fl 282 A	1 BMW 314 E	7	Flettner	7	Flettner 0007 for flight tests and familiarization
Fl 282 B	1 BMW 314 E	10	Luftwaffe	0010	service trials
Fl 282 B	1 BMW 314 E	11	Flettner	0011	reserve aircraft for service trials*
Fl 282 B	1 BMW 314 E	12	Flettner	0012	radio test flights
Fl 282 B	1 BMW 314 E	13	Flettner	0013	engine endurance tests
Fl 282 B	1 BMW 314 E	14	Flettner	0014	calibration flights
Fl 282 B	1 BMW 314 E	15	Flettner	0015	flight tests
Fl 282 B	1 BMW 314 E	16	Flettner	0016	unknown
Fl 282 B	1 BMW 314 E	17	Flettner	0017	unknown
Fl 282 B	1 BMW 314 E	18	Flettner	0018	unknown
Fl 282 B	1 BMW 314 E	19	Flettner	0019	unknown
Fl 282 B	1 BMW 314 E	20	Flettner	0020	unknown
Fl 282 A	1 BMW 314 E	1	Flettner	0001	non-flying, transmission test-bed, engine removed
Fl 282 A	1 BMW 314 E	2	Flettner	0002	non-flying, transmission and engine for B-version
Fl 282 A	1 BMW 314 E	3	Flettner	0003	non-flying, transmission and engine for B-version
Fl 282 A	1 BMW 314 E	4	Flettner	0004	non-flying, transmission and engine for B-version
Fl 262 C	1 BMW 314 E	8	Flettner	0008	transmission test-bed, new direction of rotation
Fl 282 C	1 BMW 314 E	9	Flettner	0009	new direction of rotation, general flight tests

Note: *The designations A, B and C were assigned arbitrarily, the "Specification of the B-0 and B-1 Series" was created with the Fl 282 V12. Flettner also assigned each aircraft a company Werknummer beginning with the Fl 282 V1 which was number 1585.*

Fl 282 V6, GF+YF, with a landing light mounted under the instrument panel.

10. Technical Data: (a) construction group: H 3; (b) maximum allowable flying weight: 1,000 kg; (c) wing loading: 8.84 kg/m² (based on rotor surface area); (d) maximum allowable speeds: as helicopter: forward: 80 kph, backward: 30 kph, sideways: 20 kph; as autogiro: 60 kph; (e) minimum allowable speeds: as helicopter: no limitations; as autogiro: 40 kph, aerobatics forbidden!

11. Performance:

(a) Speed: Speed is temporarily limited to 80 kph in horizontal flight in helicopter mode for reasons of structural strength, 60 kph in autogiro mode.

(b) Rate of Climb:

Height m	Rate of Climb m/sec
0	4.5
1000	3.5
1500	3.0

Maximum altitude temporarily restricted to 1,500 meters for flight safety reasons.

(c) Takeoff and Landing Performance: Takeoff and landing are made vertically. Autorotation landing in event of engine failure. Landing distance from height of 20 m: 50 m, stopping distance 15 m (with wind speed of 5 m/sec).

12. Range: Endurance of 2 hours 5 minutes at maximum allowable speed of 80 kph and a range of 168 km. The "Technical Description Fl 282 B-0 and B-1" presented above was prepared with the Fl 282 V12.

Known Fl 282 Prototypes

The following is an attempt to provide an accurate number of Fl 282 prototypes which flew or were used as test-beds based solely on authentic sources and their codes. Previous publications have quoted figures from 23 to more than 30. The sources used in this compilation:

(1) Weekly reports, monthly reports and mission summaries of the E-Stelle Travemünde (complete series not available).

(2) Delivery Plan No. 222/2 of 1/11/42.

(3) "Current State of `Kolibris' with Staffel 3/196" (11/5/44).

(4) "Summary of the Flight Times, Takeoffs and Running Times of the Fl 282" (original Flettner document from 6/7/44).

(5) Extracts from the flight logs of test pilots
Dipl.Ing. C. Bode (Focke-Achgelis)
Dipl.Ing. G. Geike (E-Stelle Travemünde)
Dipl.Ing. H. Fischer

In addition to Werk-Nummern allocated by the RLM (282000001 etc.), Flettner used internal construction numbers, such as for the Fl 265 V1 - V6 (W.Nr. 1579 - 1584) and Fl 282 v1 (W.Nr. 1585).

The following supplementary type designations are found in Delivery Plan No. 222/2:

V1 with V4 and V7	Fl 282 A
V5, V6 to V20	Fl 282 B
V9 and V9	Fl 282 C

These may be arbitrary designations, for such designations do not appear in the Travemünde documents. Furthermore, the V1, V4 and V8 were test-beds which never flew.

There never was a Fl 282 U, to which reference will be made in the subsequent chapter on the Kriegsmarine. The Navy chose the "U" for "Umbau" (conversion) or because this development of the Fl 282 was intended for use aboard U-boats.

The Flettner Fl 282 Kolibri as a 1:72 Scale Model

by Karl Kössler

Do you have plenty of patience, sensitive fingers and a good set of fine needle files? Then the model of the Fl 282 helicopter is just the right one for you. I don't know what I admire more, the technical sophistication with which Huma produced the model, or the courage with which they approached this exacting and somewhat difficult to build model. It is certainly a technical masterpiece, the large number of small and tiny parts are produced in such a way that even complicated small components, for example the two gears of the synchronizer device for the two rotors, are fully usable. It goes without saying that some cleaning up is necessary in such cases, therefore the previously mentioned needle files.

This photo provides a view of the instrument panel and rotor head with transmission. (All photographs of models by Dipl.Ing. Karl Kössler.)

After the fuselage has dried and the seams have been filled and sanded, turn to the horizontal stabilizer (29), the fairing beneath it (30), and the vertical stabilizer (31). The leading edge of the horizontal stabilizer should be less pointed; furthermore the span of the stabilizer should be only 32 mm, which is easy to correct. The fin and rudder can now be trimmed and fitted. I removed the rudder and attached it to the fin by two thin plastic strips at the attachment points as on the real thing. Care must be taken to ensure the proper angle is achieved between the horizontal and vertical stabilizers and the fuselage (the horizontal stabilizer assembly is to be modified in the future).

As with most nosewheel aircraft, our helicopter will not stand on its three wheels without the addition of some lead. There is only one place to put the weight: in the forward part of the engine compartment, right behind the cockpit wall (24). I used crazy glue to affix a strip of lead (3 mm wide, 1.5 mm thick and 12 mm long) there, which I bent to match the curvature of the fuselage.

Before installing the cockpit wall, cement the seat between the two bulges. I cleaned up the seat and cockpit wall with the help of a dentist's drill bit.

The seat needs some sort of lower support to achieve the proper angle. I achieved this by cementing a 7-mm-long propeller spinner (from the spares box) to the front of the gearbox housing, which gave the wall the necessary support (see sketch).

The many small pieces that make up the forward fuselage and pilot seat can be trimmed with a knife or file and installed. The undercarriage can also now be prepared for installation.

It is at this point that one must decide whether to build the model as it is or add details more like the original. I decided on the latter path without going to extremes; as well I lacked the time. An examination of photos 17a and 17b in the article by Herr Mohr will reveal that the fuselage frame had considerably more members than are evident on the model. This is in no way intended as a criticism of Huma, to whom I feel gratitude and admiration for producing out this kit. The necessary replacement parts are simply impossible to produce to scale.

Sprue stretched over a candle flame is readily available and is a suitable material for making the additional frame members. This method also has the advantage of allowing the modeler to achieve the desired thickness of material, or at least come close.

I first replaced Part 33 with one made from three small pieces of material 0.3 mm thick. Another piece was attached in the center, extending back to the tail fin (as on the original!). The rear undercarriage V-strut attachment point also required supporting struts, 4 in total. Two of these went to the side, to the lower fuselage panels, and two forward to the outer attachment points of Part 33.

Only when these additional struts have been added (fast-drying glue is most suitable for affixing these small parts) can the main undercarriage be installed. Again care must be used to ensure the correct angle is achieved. To this end I had to shorten one of the undercarriage shock-absorber struts (Parts 36 and 37) by about 1/2 of a millimeter.

I recommended removal of the nosewheel scissors link (small triangle at top front of nosewheel leg). Fashion a new link from two pieces of thin plastic and position it at the same height as the old one but behind the nosewheel leg. The model's appearance will be further improved by fashioning a new nosewheel leg bracing strut from stretched sprue.

Several additional struts of 0.3 or 0.5 mm thick material are necessary in the framework of the forward fuselage in order to achieve the desired overall stiffness. Once again the necessary details may be seen in photo 17b.

Yes, and then the upper part of the transmission (Part 9)! I would not recommend to anyone using this part as is. This component's actual appearance may be seen in the drawing of the gearbox and in photo 17a. Fashioning a new part is time-consuming but worth the effort. Part of this component is the two thin rods extending from behind the cockpit wall up to the steering mechanism. Another part that is absent is the control bridge, which joins the two elements of the rotor heads identified as "directional indicators." It is easy to make. One should also at least suggest the presence of the servo drive behind the actual rotor transmission.

There is not much to say about finishing

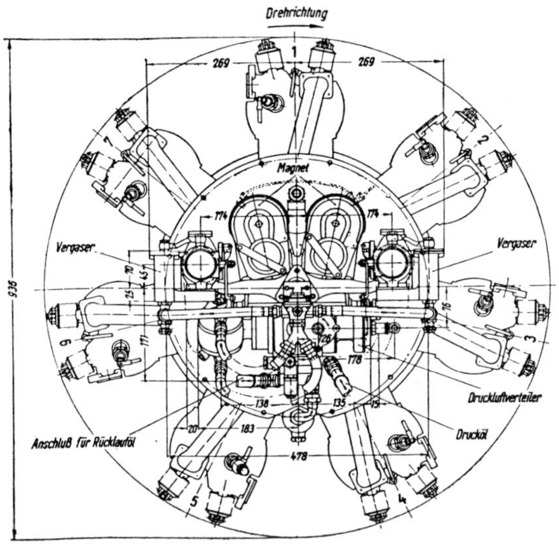

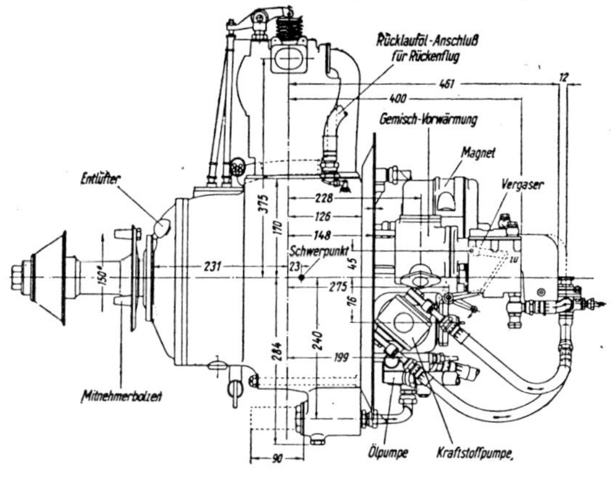

Left and above: Front and side views of the Sh 14 A (BMW 314 E) engine.

This view provides a good view of the helicopter's drooping rotor blades.

the model – best done by airbrush. As the drawing indicates, all upper surfaces were sprayed Dunkelgrün 71, the undersides Hellblau 65. The pilot seat and forward fuselage framework – at least the inner part – were probably left in Schwarzgrau 66.

As I wished to prevent paint from running into the rotor shaft holes during spraying, I covered the projecting rotor tubes and the "directional indicators" beneath them with Maskol. Some of the masking fluid seeped into the shaft hole on one side. Removing it afterward proved almost impossible. My attempts to make the shaft movable again must have resulted in several of the gear teeth breaking off, in any case the rotors are synchronized for only part of their range of travel – a pity. I therefore recommend using some softened wax or plasticene to cover the rotor shafts and their mounting holes during spraying.

The rotors, which are painted after being cleaned up and glued to the damping disc, are of course the last components to be added. Note that the right rotor – seen from above – turns clockwise, and the left counterclockwise. If the left rotor is aligned with the fuselage, the right should be exactly perpendicular to it (see photo left). However both are parallel to one another in the 45 degree position. The rotor blades should droop slightly for a natural appearance, which can be achieved by bending the blades gently by hand.

In conclusion a word about applying the decals. It is important that they are allowed to soak thoroughly and stick well. I used the well-known method of eliminating the inevitable decal film: first spraying the entire area of the decal with clear gloss varnish then overspraying with matt varnish after the decal has dried.

The Technische Amt and the Fl 282

It was not until mid-1940 that the Technische Amt (RLM Technical Department) became more deeply involved with helicopter development, and then at the urging of the OKM. Even so, it did not assign the work a special priority level. The following are several extracts from the so-called "GL" and "Department Head Conferences" which were chaired by Generalluftzeugmeister (Chief of Air Armaments) Milch. The main purpose of the latter was to examine current programs by airframe and engine manufacturers and their feasibility. The selected extracts refer only to the Fl 282.

GL Conference GL 1/Genst.6.Abt.

17/5/40: Genst.6.Abt. was informed that the GL cannot yet accept responsibility for an early start-up of production of the Fl 282. A request was made that the increase from 20 to 30 0-Series aircraft be delayed. The Genst. expressed the opinion that Flettner offers no guarantee for a quick and on-time start-up of production. Therefore, in view of this type's importance, the Genst. requests an investigation as to whether it can be brought to the production stage and built by another firm.

24/5/40: The O-Series is increased from 20 to 30 aircraft.

14/8/41: Genst. requests 10 Fl 282 and 30 Fa 330 aircraft per month for use by the navy.

This view from above shows that two blades were attached to each rotor head.

Department Head Conference

19/6/42: Concerning helicopters, Director Freytag pointed out that it would be advantageous to divide the Flettner operation into two parts, if the Fl 282 ever reaches the production stage. Flettner is a design firm and not suitable for quantity production. Construction of prototypes should continue in Johannisthal under the direction of Flettner, while series production would be handed over to his colleague Glauner.

19/7/42: Sellschopp: Then come the helicopters, the Fl 282 and the Fa 330, one at two machines per month, the Fa 223 at five per month.

Milch: Has development already reached the stage where they can enter production? I consider these small numbers an experimental series. And then there is the BMW 314 engine, what is going on with it? They're building only half an engine each month. What will the Flettner be used for?

Friebel: It is the aircraft that's supposed to be employed in cooperation with the navy.

Milch: What is it supposed to do?

Friebel: It's supposed to be able to take off from small warships.

Kleinrath: The Fl 282 is contained in the naval command's request.

Frydag: Land-based tests are currently under way and they look very convincing. I recently made an inspection of Flettner. The firm's main problems are: Flettner is an inventor and built his rotor transmission himself. I said to him: "You'll get nowhere the way you're going. You aren't even set up to produce five transmissions a month." He is very well set up for building prototypes and the individual transmissions are also very good. But he fights tooth and nail against giving a license for the things to a gear or transmission factory. He maintains they can't do it. I

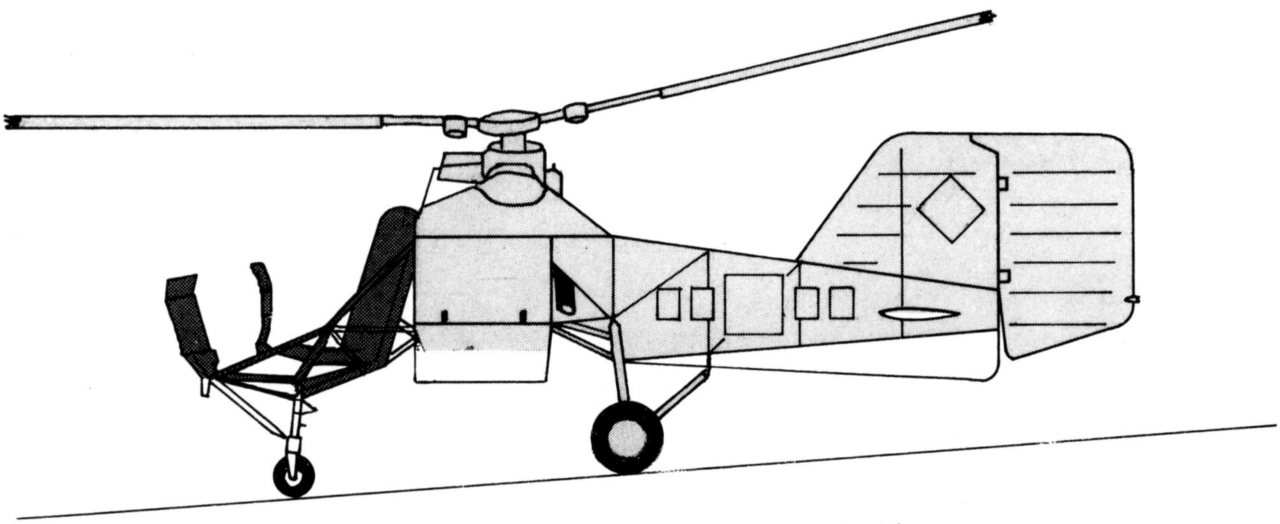

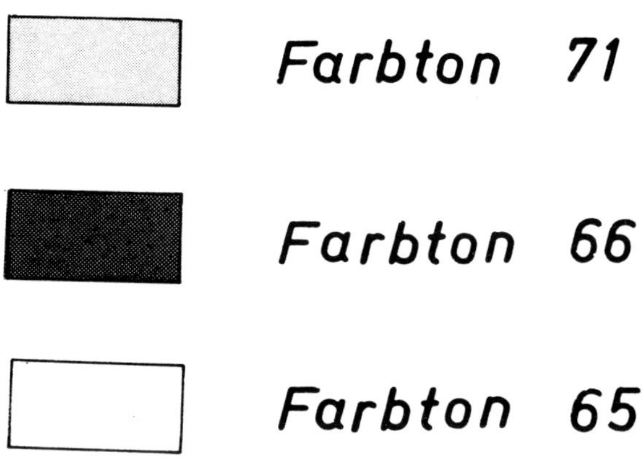

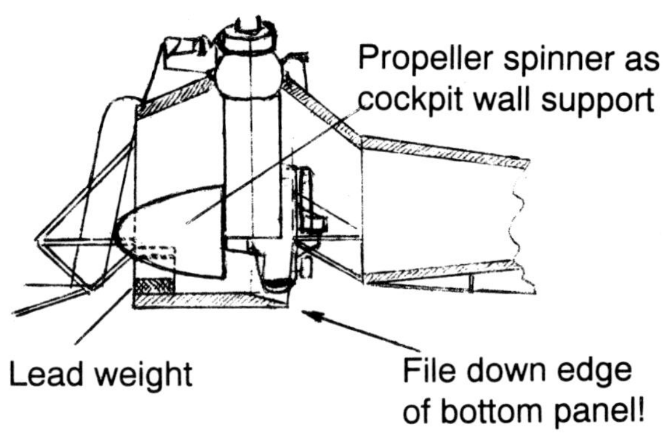

The Fl 282 V12 CJ+SF with partial plexiglass enclosure of the cockpit. (All photos from the Mohr Collection)

was unable to convince him, perhaps he'll have to be forced into it. He is supposed to build 1 1/2 units a month and barely completes them. Flettner is much too poorly supported to handle this. He does an excellent job on the prototypes but will never succeed in series production.

Milch: The aircraft were supposed to come at the end of 1942 and its already the middle of 1944. That's a period of a year and six months, not even 1 1/2 machines per month.

Frydag: I told him: "You haven't finished the Fl 282 yet." But he's an inventor: he has something new and then he makes something new. If the Fl 282 should really be needed, we could divide the Flettner operation into two parts: leave Flettner and the small operation in Johannisthal to build prototypes and take away the series production and give it to his factory chief, Herr Glauner.

The Fl 282 is a free-flying helicopter. It comes down and then they grab the thing at this height (indicates about 2 meters), attach the rope and pull it through a pulley. This is how they get the thing down even when the ship is tossing. (Ed. note: Frydag's rather inexpert description is meant to portray the deck-landing process.)

Vorwald: "Wagtail"? But that's a tethered helicopter.

Frydag: Flettner has one too. Quite a simple thing, no rear fuselage, a square box, no tail surfaces or anything. (Ed note: preliminary experiment with the Fl 282 V9 for the use of the Fl 282 on large U-boats).

Milch: Does anyone sit in it or is it done by remote control?

16/1/43: Kleinrath: In regard to helicopters, I must report that the army is now requesting 1,000 Fl 282s. Based on the first documents from GL/C we have replied that the questions of production capacity, workers, and so on, including aviation fuel, must be handled by the army. It is strictly a one-time request.

Pasewaldt: That is an absolute exaggeration, for there's no use for such numbers of this aircraft. The possible operational uses of this "circus aircraft" are so minimal that it would be a crime if we were to make 1,000 of them. (Ed. note: When he used the term "circus aircraft" Dr. Pasewaldt was probably thinking of the public flight demonstration of the Fw 61 V1 D-EBVU Deutschland by Hanna Reitsch in the Deutschlandhalle in Berlin in February and March 1938).

The Army Sees the Helicopter as the Solution to all its Problems

Kleinrath: Yesterday I informed Oberst Vorwald that the Fl 282 which the GL made available to General Zeitzler was mistakenly on view at the big display of winter vehicles for the army. The army is now acting as if this helicopter were the solution to all its problems. In any case we have written o the army that there can be no doubt that everything they are asking of us can only be achieved with the army's help: workers, raw materials and fuel.

Vorwald: We won't let the thing begin until we receive a clear request from the army as to how many Fl 282s we are supposed to plan for them.

Kleinrath: Yes, but with the influence now exerted by the Minister of Armaments and Munitions (Speer) and the gentlemen of the army acting as if it's their crate, we fear that the Fl 282 will be demonstrated to the Führer. There's reason to fear that that could result in a sudden order from the Führer to build 1,000 Fl 282s for the army. In my opinion the matter should be taken up with the Reichsmarschall so that he can oppose this.

29/2/44: Vorwald: Cancellation of the Focke-Achgelis and Flettner helicopters and use of the capacity for the fighter firms.

Kindling: The helicopter companies were already significantly restricted in autumn 1943. The Flettner company in Schweidnitz now has only 12 designers and 88 factory personnel plus 20 auxiliary workers, or 120 men in total.

Hermann: A suggestion for Schweidnitz would be to leave the block (Flettner) together as a working group and build the Me 262 there.

Kindling: There's only a single hangar in Schweidnitz, half of which is used for production and the other half as an aircraft hangar.

Milch: Stop everything there! We'll take over this hangar!

Kindling: The airfield is very small, so that we would like to propose carrying out modifications and so on there. In any case the airfield is too small for building complete fighters. We need to decide whether helicopter development should be completely shut down.

Milch: Shut it down temporarily! Work can resume after the war. Professor Focke should go and work in an institute somewhere! All personnel will be assigned more important tasks.

Descriptions of Persons:

Milch, Erhard
Generalfeldmarschall and State Secretary for Aviation, succeeded Ernst Udet as Generalluftzeugmeister (Director General of Air Armaments, abb. GL) in Nov. 1941

Friebel Engineer GL Planning Department

Frydag, Carl
Director on Reichsmarschall's Industry Advisory Council (Aviation)

Hermann, Rudolf
Engineer-General. Development planning in the Luftwaffe General Staff

Kleinrath, Kurt
Engineer-Colonel. Head of Luftwaffe General Staff Department 6 (Equipment)

Kindling Engineer with GL/C production

Pasewaldt, Georg
Doctor of Law. Head of the GL/C-E (Development) section

Sellschopp, Heinrich
Engineer-Colonel. GL Planning Department

Vorwald, Wolfgang Engineer, General Staff Colonel. Head of the RLM's Technical Department

Zeitzler, Kurt
Generaloberst. Head of the Army General Staff.

Known Fl 282 Prototypes

	Prototype Code	1st Entry in Log Book	Use
V 1		15/08/41	transmission test-bed until 21/11/41, then set aside by company, engine and transmission removed
V 2	30/10/41		first flight by Hoffmann on 30/10/41, fully-glazed cockpit, resumed test flying after various modifications (March 1942 to 22/5/42 last flight), set aside by Flettner, engine and transmission removed
V 3	GF+YC	02/12/41	fully-glazed cockpit, enlarged vertical fin, horizontal stabilizer with end plates. Two altitude flights by Hoffmann on 27/4/42 to 3,500 and 3,800 meters (in 36 minutes) over takeoff point
V 4			reserve test-bed for V 1
V 5	GF+YE	08/01/42	Deck trials on the Greif in Travemünde beginning in Oct. 1942. Resumption of training at E-Stelle in Jan. 1943 until 15/9/43. Arrived at Flettner on 24/11/43 after overhaul by Flettner, equipped for land/sea trials
V 6	GF+YF	11/05/42	deck trials on the Greif, service trials aboard the Drache in the Mediterranean by E.-u.L.Kdo.20 from Oct. 1942 to Jan. 1943. Total loss on 10/5/43 35 miles NE of Christiansöye while being flown by Hptm. C. von Winterfeldt
V 7	CJ+SA	22/05/42	prepared by Flettner for flight tests and familiarization of new pilots. To E-Stelle in Jan. 1943 for training until 26/8/43
V 8		26/06/42	test-bed for reversed direction of rotation of rotor blades
V 9		15/01/43	flight tests by Flettner with new rotor direction of rotation, fuselage removed aft of firewall, until 19/3/43
V 10	CJ+SD	20/09/42	Oct. 1942 to Jan. 1943 service trials on the Drache together with the V6, subsequently sent to Flettner for overhaul. Flown to E-Stelle on 11/9/43, input checks. 11/5/44 with 3/196, flew with rotors of the V 14
V 11	CJ+SE	03/11/42	planned as reserve aircraft for service trials, used for training until 26/11/43
V 12	CJ+SF	12/12/42	collected from Flettner by BAL, subsequently used for gust measurements J./F with Flettner, familiarization of E-Stelle pilots Dumke and Fischer 9/4/43 flown to Travemünde by Dumke April 1943 tests of FuG 19 (ultra-short wave radio) 28/5/43 2 hour and 53 minute flight to Berlin for overhaul (Flettner 19/9 to 25/9/43 deck-landing trials on the Greif off Travemünde Oct. 1943 damage of ribs at rotor attachment point, rotors changed, blades for V 5 (land/sea trials)21/11 to 27/11/43 cold start trials 12/12 to 22/12/43 conclusion of cold start trials, equipped for dropping of stores (2 x Fl.W.B.5 5-kilo anti-submarine bombs) 23/12/43 creation of "Interim Type Sheet" for Fl 282 B-0 and B-1 March 1944 installation of prototype bomb-dropping system (Vemag 2 Fl.W.B.5) complete April 1944 shipped by rail to Schweidnitz for general overhaul June 1944 still with Flettner, total of 412 flights
V 13		02/03/43	planned for engine duration trials with Flettner, including reverse flight tests with empennage removed
V 14	CJ+SH		planned for calibration flights, conversion June 1943, on 11/5/44 with 3/196 without rotors, unserviceable
V 15	CJ+SI	15/04/44	flights with reversed direction of rotor rotation, believed to have been captured by the Americans (FE-4614?)
V 16		23/08/43	no information
V 17	CJ+SK	01/06/43	Dec. 1943 input checks at E-Stelle, failure of tether harness during engine run-up. Use of all tether harnesses forbidden pending strengthening of center-section. Deck-landing trials on a platform (16 m²) on the Lufthansa Bridge in the Pötenitzer Wiek began on 9/2/44 Crash with new landing harness on this platform, pilot Dipl.Ing. Geike of the E-Stelle (see photo Page 58)
V 18	CJ+SL	30/07/43	11/5/44 with 3/196, handed over to the E-Stelle for further tests
V 19		28/09/43	reverse flight tests with empennage removed
V 20	CJ+SN	26/11/43	11/5/44 with 3/196, Pillau, unserviceable
V 21	CI+TU	04/05/44	two-seater, top Schweidnitz in June 1944 for partial overhaul, then transported to Travemünde by rail
V 22			no information
V 23	CI+TW		two-seater, wooden cockpit side walls, V-shaped horizontal stabilizer, captured by the Americans, FE-4613and T2-4613.

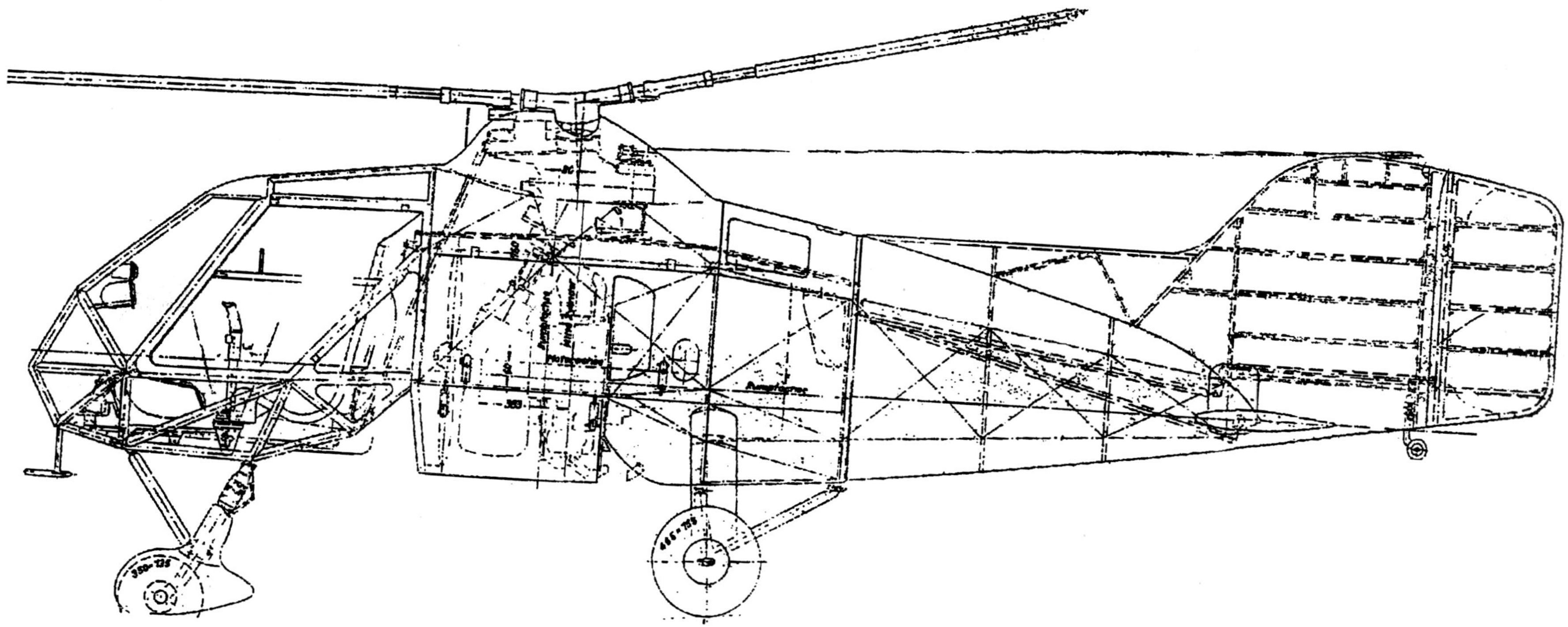

Original factory drawing:
Flettner Fl 282 with fully-glazed cockpit.

Drawings by Ralf Swoboda

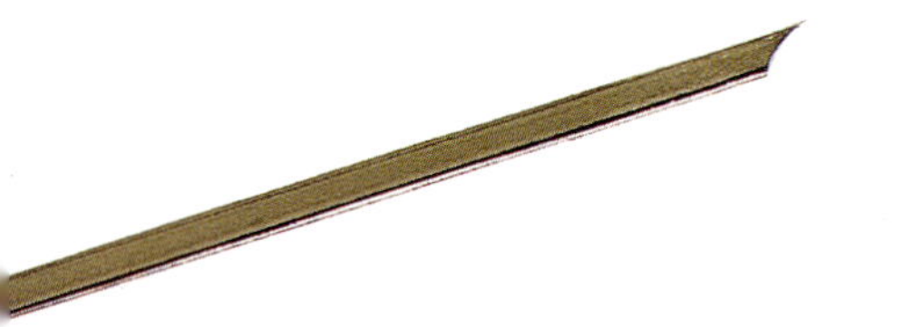

CI TU

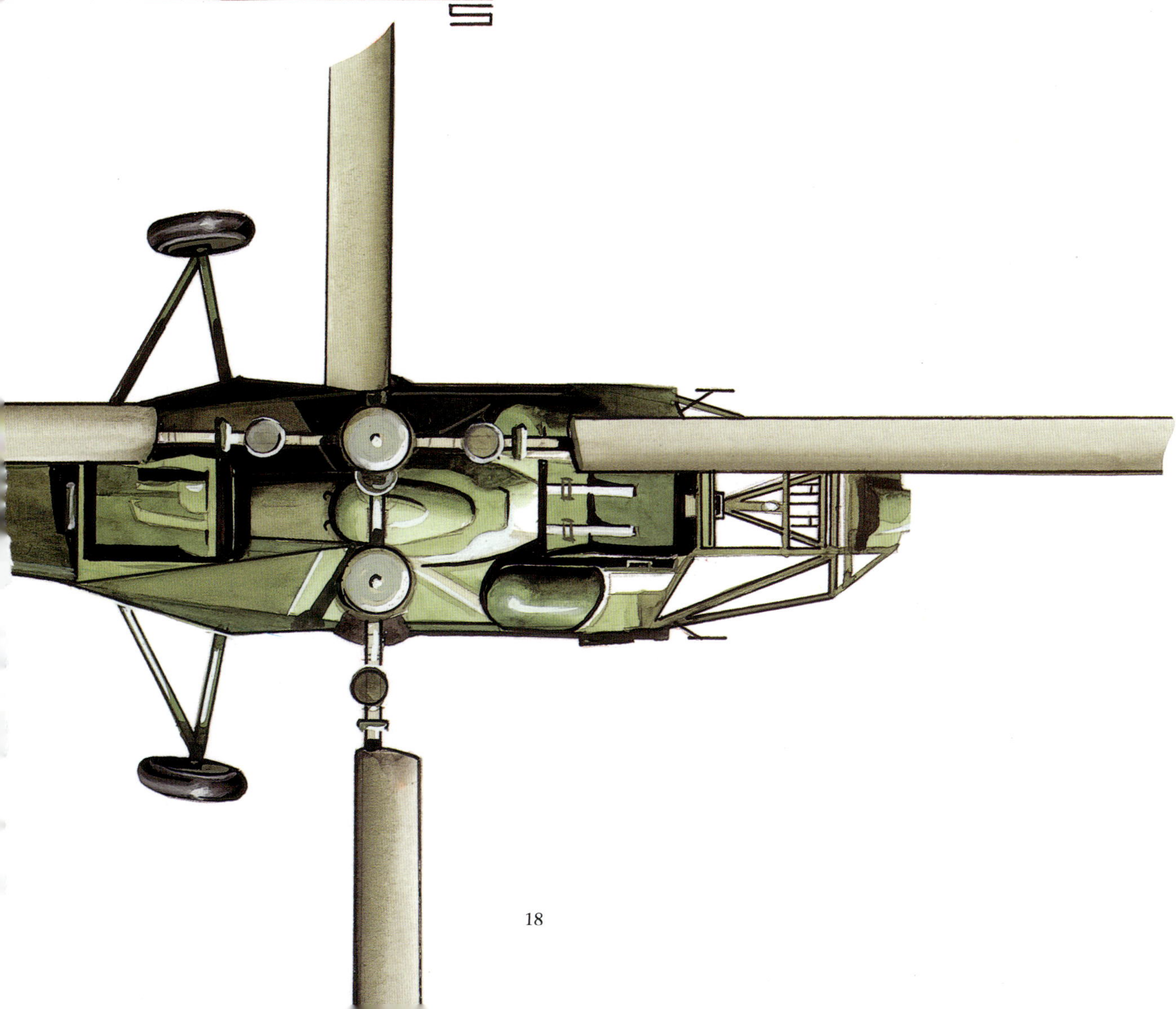

Original factory drawing:
Flettner Fl 282 with fully-glazed cockpit.

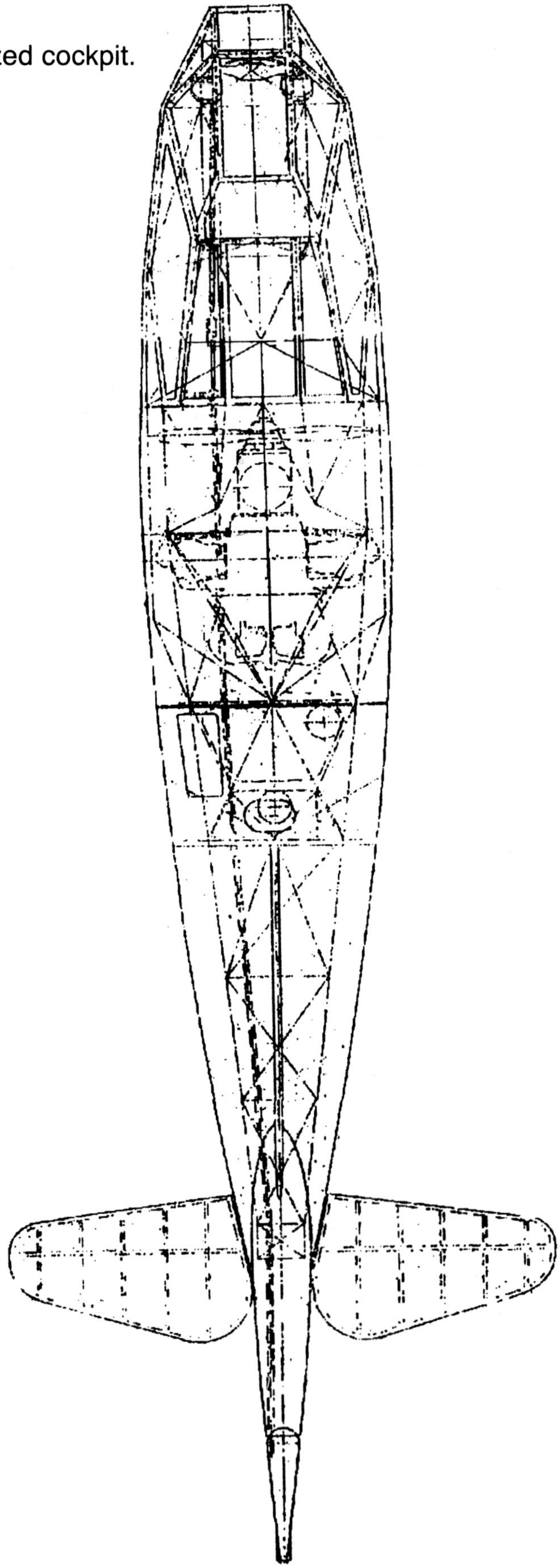

Instrumentation in the partial cockpit glazing (from left to right:) vertical speed indicator, airspeed indicator, turn-and-bank indicator, rotor RPM indicator, rotor blade angle indicator. Below: master compass, above two instrument lamps.

Port instrument console (from above): altimeter, double pressure gauge for fuel and oil, double oil temperature gauge (upper and lower transmissions), double motor oil indicator (inlet and outlet), ignition switch.

Port side of the cockpit of the Fl 282 V12: FuG 19 (top), throttle and blade incidence levers, tether rope release lever, trim lever (beside the FuG 19). The pilot's view from the semi-open cockpit was outstanding.

Summary of Flight Times, Takeoffs and Running Times of the Fl 282. (As of 28/6/44) From aircraft log.

Aircraft No.	Flight Time*	Number of Takeoffs	Running Time**	Aircraft Log Entry First-Last		Remarks
V1	Used as test-bed		125 hr. 39 min.	15/08/41	21/11/41	
V2	6 hrs. 29 min.	17	25 hr. 6 min.	30/10/41	22/5/42	First free flight by a Fl 282 on 30/10/41
V3	18 hr. 39 min.	60	207 hr. 40 min.	02/12/41	31/01/44	
V4	reserve test-bed for V1					
V5	41 hr. 10 min.	255	76 hr. 14 min.	08/01/42	15/09/43	reverse flight tests with aft fuselage removed
V6	76 hr. 31 min.	174	102 hr. 18 min.	11/05/42	10/05/43	100% writeoff by Hptm. C. von Winterfeldt
V7	46 hr. 56 min.	558	110 hr. 09 min.	22/05/42	26/08/43	
V8	used as test-bed		77 hr. 40 min.	26/06/42	24/09/42	tests with reversed direction of rotor rotation
V9	50 min.	18	10 hr. 41 min.	15/01/43	19/03/43	flights with reversed direc tion of rotor rotation, aft fu selage removed
V10	82 hr. 03 min.	372	103 hr. 49 min.	20/09/42	28/06/44	3/196
V11	78 hr. 53 min.	303	90 hr. 19 min.	03/11/42	26/11/43	officially with 3/196, but still with manufacturer at the time
V12	90 hr	412	101 hr. 54 min	12/12/42	28/06/44	
V13	100 hr. 53 min.	215	142 hr. 44 min.	02/03/43	02/03/44	reverse flight tests with aft fuselage removed
V14	28 hr. 32 min.	116	46 hr. 59 min.	?	27/06/44	3/196
V15	14 min.	3	1 hr. 19 min.	31/03/44	15/04/44	flights with reversed direction of rotor rotation
V16	47 hr. 21 min.	108	59 hr. 51 min.	23/08/43	26/06/44	
V17	24 hr. 02 min.	168	33 hr. 26 min.	01/06/43	13/04/44	100% writeoff by Dipl-Ing. Geike E-Stelle Travemünde (Photos 17a and 17b)
V18	20 hr.	80	33 hr. 35 min.	30/07/43	28/06/44	3/196
V19	10 hr. 38 min.	39	21 hr. 36 min.	28/09/43	25/06/44	reverse flight tests with em pennage removed
V20	37 hr. 18 min.	158	37 hr. 33 min.	26/11/43	10/06/44	3/196
V21	—	—	5 hr. 10 min.	04/05/44	06/05/44	two-seater, no free flights as yet
Total	710 hr. 29 min.	3956	1423 hr. 42 min.			

* = free flights only with no tethered training flights

*** = all test runs + tethered flights + free flights + warm-up times*

The Production of 1,000 Fl 282s by BMW

As stated in the introduction, all previous authors to concern themselves with the Fl 282 have made passing reference to the above "fact" without offering any details. The source of the story of the 1,000 Kolibris appears to have been an equally brief note in "B.I.O.S. Overall Report Number 8," which spoke of "a batch of one-thousand Fl 282s" and the awarding of a contract to the BMW factories in Munich and Eisenach. The report even went so far as to state that the jigs and tools were under preparation.

For two years the author had the opportunity to sift through the entire contents of the old BMW archive, including the records of Eisenach branch plant, while conducting supplemental research for the book BMW – eine Jahrhundertgeschichte (Part One) by H. Mönnich, specifically the section concerning BMW power plants and aircraft with BMW engines. Not a single document concerning the 1,000 Kolibris turned up.

Even if nothing could be found of this large and – for an aero engine factory – unusual production contract for 1,000 helicopters, it should have been possible to at least find a corresponding RLM contract or some reference to the 1,000 BMW 314 E engines required for the construction of the Fl 282. (The Sh 14 A which powered the Fl 282 was a BMW 314 with a special transmission.) Production of the Sh 14 A had ended in the summer of 1939. Once again nothing was found.

A study of the RLM's C-Amt Program and the GL conferences offers some information. The Aircraft Program (Prototypes) Delivery Plan No. 221/1 of September 1, 1942 lists the total number of Fl 282 Bs delivered by Flettner at 16 examples. Fl 282 B was the designation of the production version, however these 16 aircraft were built as prototypes.

The next delivery plan (No. 223/1) of April 15, 1943 presents a different picture. The program once again shows 16 Fl 282s for Flettner. Additions to the delivery plan were: BMWE (which stands for BMW Eisenach) joining Flettner as manufacturer with plans to build 150 examples of the Fl 282 B, the first to be delivered in January 1944. As well approval was given for the procurement of materials for another 500 aircraft.

The author also discovered the date on which this large order first appeared in an aircraft program, something that all previous authors have failed to do. The request for 1,000 Fl 282 helicopters was submitted to the GL by the army in January 1943 but was paid little heed by the RLM. If the 1,000 Kolibris had been built, it would have been for the Army and not the Luftwaffe!

The next aircraft program (No. 224/1) of October 1, 1943, or six months later, stated (referring to the Fl 282): "All approvals for the release of materials and production preparations issued prior to the publication of this delivery plan are hereby revoked."

Program 226/1 of May 15, 1944 canceled the delivery of Fl 282s, which had been reduced from 16 to 7 examples. This was the ultimate end of the Kolibri, the production of prototypes having been halted the previous November.

Without delving any deeper into the feasibility of the aircraft program as devised by the Luftwaffe command – interested readers may refer to Die Deutsche Luftwaffenführung 1935-1945 by H. Boog – there are several other points to consider regarding the plan to involve BMW in the production of helicopters. Leaving aside the shortages in materials and labor which were already beginning to make themselves felt in 1943, the BMW aero-engine factory in Munich couldn't possibly become involved in the production of aircraft as it lacked the required space and production capacity. Furthermore, production of them BMW 801 for the fighter program and work on the BMW 003 jet engine had absolute priority. The Eisenach factory built the BMW 132 A/T and, from 1942, engine parts for the BMW 801 and BMW 323. Production of parts for the BMW 003 engine was added at the end of 1943.

This, in rough outlines, is the story of the "1,000 Kolibris by BMW," of which only 150 existed on paper in the form of an aircraft delivery plan. Claims that preparations were under way at BMW for production of the Fl 282 are therefore unfounded.

The Kriegsmarine naval staff, together with the U-boat command group (1./Skl.U/III), recognized the advantages of a helicopter as a shipboard aircraft early on. Even before the war much thought was given to the problem of its use on various types of warships (including submarines). In addition to autogiros and small airships, these early deliberations also gave consideration to the use of Fi 156 Storch liaison aircraft, which had carried out initial deck-landing trials on the air-traffic safety ship Greif in 1937. These considerations were put down

Photos 17a and 17b: Fl 282 V17, CJ+SK, following a crash-landing at Travemünde. Note the rotor heads—intermediate link with damper—and Holman end fork with rotor blade. Also clearly visible are the instrument panel, the yellow fuel triangle (87 octane) over the filler point and the code letters with Balkenkreuz under the horizontal stabilizer (bottom photo).

on paper in a study by Leutnant zur See Lindner dated July 24, 1937, titled Fixed-wing Aircraft or Helicopter. The paper made very clear the helicopter's superiority over conventional shipboard aircraft; later war service clearly demonstrated that the fixed-wing shipboard aircraft was a "weather-limited weapon of opportunity." The report's concluding paragraph is interesting:

"We fail to appreciate the significance of the introduction of helicopters on a larger scale by the U.S. Navy (Kellet KD-1A) and the founding of an academy for rotary-wing matters in the U.S.A., for this type has been under test for years and its use is proof of the positive results of these tests. If this move by the U.S. Navy gives us cause for thought, then the technical advantage that Germany has achieved through the building of helicopters may spur us to attack this question as quickly as possible, in order to use our technical advantage to assure a tactical one as well."

The naval staff's evaluation of the study included the following paragraph:

"We also share his (Lindner's) fear that we might fall behind the Americans. Therefore the request to the general staff (of the Luftwaffe) that we now approach the problem energetically from a military standpoint." (17)

CJ+SF was demonstrated to high-ranking Italian officers at Rechlin on 10/1/1943. Fourth from left is Anton Flettner. (All photos from the Birkholz Collection)

The naval staff established a preliminary list of the technical and tactical factors affecting the development of a helicopter for military purposes in a letter to the Inspectorate of Naval Aviation dated December 19, 1939. Identification of these factors was based on film of the extensive tests carried out by the navy with a Fl 265, which was shown in the RLM (10/10/39), the above mentioned study, and ideas developed by the naval staff. The proposed path of development of the Flettner helicopter projected the development of a two-seater (Fl 282) and a submarine-based helicopter (Fl 282 U). (18)

The L.In.8 issued the definitive technical-tactical factors for the development of helicopters on January 29, 1940; three different types of helicopter were described:

(1) Coastal helicopter, use in coastal areas,
(2) Shipboard helicopter, use from large warships,
(3) Small helicopter, use from submarines. (19)

In a letter dated April 19, 1940, the Kriegsmarine High Command revealed that enemy submarine operations in the North Sea and the Skagerrak had increased to such a degree that the German anti-submarine forces were having to fight hard to master the situation. In its view the Flettner helicopter that had already been demonstrated (Fl 265) was a most suitable means of addressing the problem. According to information from the LC 2, the Flettner helicopter was still in the development and testing phase and by summer 1940 would reach the stage where individual examples could be made available. (20)

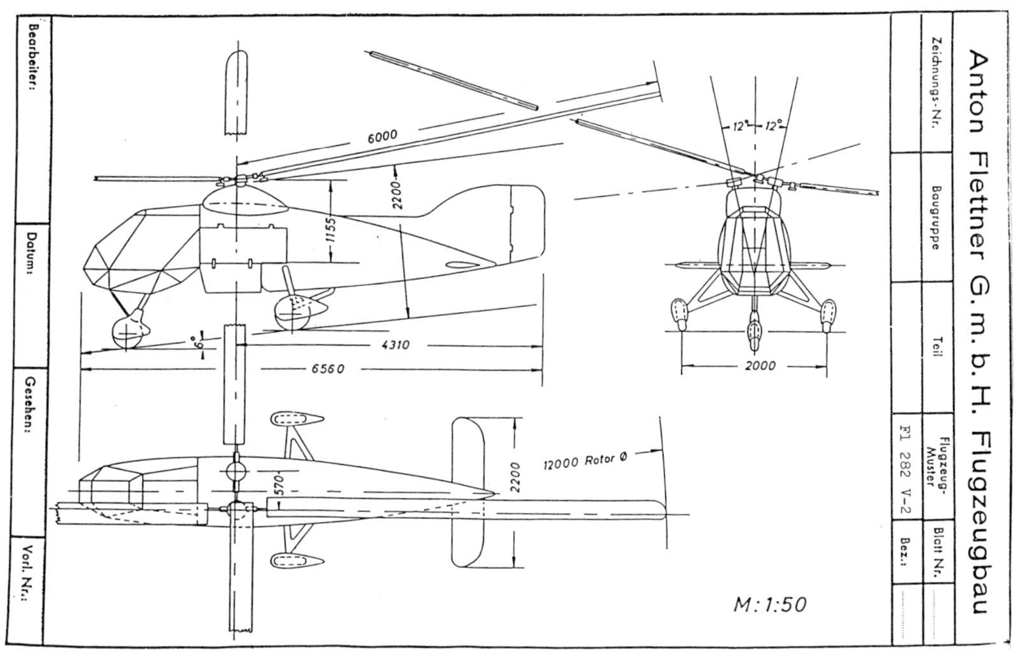

Three-view drawing of the Fl 282 V2 as presented to the Naval Staff in December 1940. (Mohr Collection)

A report on the state of development of the Flettner helicopter released on June 24, 1940 revealed that this information from the GL/LC 2 was too optimistic:

Fl 265 V1 and V2 in company trials (V3 written off on 21/8/39), V4, V5 and V6 available for flight testing between 16/6 and 15/8/40 (no military use foreseen).

Anticipated inspection of working mockup on 8/5/40. Start of flight testing with the V1 beginning of September 1940.

(Ed. note: The first flight by an Fl 282 did not take place until October 30, 1941!)

At this time no sort of priority level had been established for the design and construction of helicopters! (21)

More than a year later, the Skl.U/III declared in a letter dated June 7, 1941:

"The Skl.U/III has closely monitored the development of the Flettner helicopter for the purpose of submarine defense for the last 1 3/4 years. As a result of setbacks in development of the transmission, the three helicopters which were to be allocated to submarine defense could not be delivered in summer 1940, and the company is only now at the point again where it can think about delivering helicopters for front-line purposes.

In view of this situation the Skl.U/III has again contacted the RLM and the Flettner company again in order to obtain the earliest possible delivery of helicopters. It turns out that the army is now also showing strong interest in the Flettner helicopter, mainly for use by combat engineers. The interest on the part of the Army Ordnance Office is such that the Flettner company was even promised 40 specialists to help speed subsequent development. This raises fears that the Flettner company might be forced to work mainly for the army.

Such a course of action would be all the more detrimental to the navy because in addition to the anti-submarine role, the helicopter is being considered as a general reconnaissance aircraft as well. The reasons which appear to make the Fl 282 especially suitable as a shipboard reconnaissance aircraft are as follows:

(1) The aircraft takes up very little space. With the rotor blades removed, the Type 282 has a fuselage length of 6.56 meters and a width including undercarriage of 2.40 meters. By parking in ranks, 10 helicopters can be put in the space required by a conventional shipboard aircraft.

(2) Preparations for flight take only a few minutes. The small, light aircraft is easily rolled out of the hangar on to the ship, the rotor blades are attached quickly. Movement of the ship has very little effect on preparations for flight.

(3) The rotor power plant can run at full power while the aircraft is tethered. When the tether is released the aircraft leaps upward from the deck (jump takeoff).

(4) After taking off the helicopter can, if necessary, climb vertically to 3,500 meters. Economical cruising speed is 60 kph, high-speed cruise 120 kph, and maximum speed 150 kph, but this is uneconomical. Depending on speed, endurance is 1.5 to 3 hours.

(5) The view from the helicopter is excellent as it can hover in the air and turn to all sides.

(6) In the event of engine failure the rotor blades automatically switch to autorotation, or gliding flight, which if necessary allows the pilot to abandon the aircraft.

(7) For landing on the stern of the ship the helicopter drops a landing rope from a height of about 10 meters above the deck. The rope is passed through a pulley and held taut by hand, which allows the helicopter to match the up and down movement of the ship. In this way it can be brought down to the deck without the danger of damage.

This photo of CJ+SF provides a good view of the pilot seat with harness and the rotor head. The four "pots" on the blade roots (arrows) are vibration dampers.

(8) The ability to operate in bad weather and complete independence of sea-state conditions, which have been demonstrated on various occasions, together with the large number of helicopters that can be accommodated on deck, make it possible to maintain a constant airborne watch on the ship regardless of wind and sea conditions. The extent to which better aerial reconnaissance could be provided through the use of helicopters as opposed to conventional shipboard aircraft was illustrated by the Admiral Scheer during its most recent sortie. In 180 days at sea it was only able to use its aircraft on 28 days.

If these advantages offered by the helicopter are acknowledged, it appears necessary to immediately take all measures necessary to equip our anti-submarine vessels and other surface vessels with helicopters.

Given the existing labor situation and the low priority level assigned the Flettner company, the outlook for achieving this is extraordinarily poor."

At the same time Flettner requested 50 qualified engineers and technicians for the technical office and 200 skilled workers for the construction shop in order to meet the planned delivery dates for the Fl 282. As well, all expansion work was supposed to begin by July 1941 and the funds to be invested by the RLM were to be made available.

The following paragraph from the above-mentioned letter is noteworthy:

"At the present time Skl.U/III is negotiating with the Flettner company for a helicopter for large submarines. Unlike existing and

V12 CJ+SF during its demonstration in front of the Italian military delegation. The pilot is believed to have been Flettner company pilot Guisting. Note the code under the horizontal stabilizer. Note the angle of incidence of the horizontal stabilizer as the aircraft prepares to land.

planned machines, the shape of the helicopter envisaged for these submarines will be approximately that of a 'standing cylinder.' It will be accommodated in a pressure-tight container mounted behind the submarine's conning tower. For stowage it will be necessary to lower the helicopter, which will have folding blades, into the container by elevator. This submarine helicopter will be able to climb to 3,500 meters and reach a speed of 50 to 60 kph, which will be adequate for the submarine's requirements. Development of this helicopter type appears especially important." (22)

Contained in a memo dated June 19, 1941, the naval staff's assessment of the situation in the air armaments sector was right on the mark:

"Opposing the wish to push on and accelerate the completion of the helicopter is the requirement to cooperate with the industrial capacity of the Luftwaffe at the points of main effort. But there is no doubt that in the present military situation the Luftwaffe is not in the position to release labor forces for aircraft that will not be able to make a decisive contribution to the outcome of the war. As well the navy is not in a position to release skilled workers to Flettner to force the construction of helicopters. The naval staff therefore does not see itself in a position to request a higher priority level for the helicopter in the Luftwaffe's armaments program. It will, however, ask the Commander-in-Chief of the Luftwaffe to promote the building of the Flettner helicopter within the realm of possibility and to place the construction of helicopters in the highest 'SS' priority level." (23)

Space prevents us from quoting further naval staff files on the Fl 282, but from the extracts provided it is obvious to what degree the Kriegsmarine – in contrast to the Luftwaffe – was willing to press for the rapid development and quantity production of the Fl 282 to equip its ships. However, all its efforts were to prove fruitless.

As late as mid-1944, after the RLM had halted the construction

of all helicopters and autogiros and the "helicopter matter" had ended with the disbandment of 3/196, the naval staff presented the R.d.L.u.Ob.d.L. and GL/C-E 2 with a study dated June 15, 1944 titled: Tactical-Technical Demands on Helicopter Aircraft for Operations over Water. The study made reference to the following types:

(a) Fa 223, land-based, for convoy escort, anti-submarine warfare and control of anti-submarine units.

(b) Fa 223 Zwilling (Twin), (development Dackel – Dachshund – and Krabbe – Crab), land-based for special transport, torpedo carrier ("flying motor-torpedo boat"), and as flak carrier (with at least one 20mm quadruple flak) for convoy escort.

(c) Fl 339, ship-based, as interim shipboard aircraft beside the Ar 196 and as control aircraft for anti-submarine units in offshore areas.

All of these types were to be equipped with a marine search radar (range approximately 10 km.).

The Fl 282 in the Diary of the Luftwaffe General Attached to the Commander-in-Chief of the Navy

Entries made in the war diary of the Luftwaffe general attached to the staff of the Commander-in-Chief of the navy shed further light on the Kriegsmarine's repeated requests for the Fl 282 and the associated difficulties.

The Luftwaffe Inspectorate of Naval Aviation held a conference on April 17, 1942, in order to determine the numbers of aircraft that were being requested by the Kriegsmarine now that the Flettner Fl 282 had been declared ready for production. As a result of this examination firms were to be identified and allocated work in order to ensure production of power plants and transmissions in the required quantities.

The naval staff came up with a figure of fifty Fl 282s for the first production batch with which to equip its warships (including training ships, auxiliary cruisers, icebreakers, blockade runners, escort tankers and supply vessels), without allowing for losses, attrition and the supply quota.

The 1.Skl./U III requested an additional 40 aircraft for the anti-submarine units and a monthly quota of 5 to 6 Fl 282 U types for the large submarines. The total number of aircraft was thus 90, not counting the Fl 282 U. This request on the part of the navy exceeded the GL's greatest expectations – or fears, for the then valid Delivery anticipated a total of 16 Fl 282s. Moreover there were two production bottlenecks: the transmission and engine. No company had been as yet been selected to produce the transmission and production of the Sh 14 A (on which the Fl 282's BMW 314 engine was based) had ended in the summer of 1939. The only remaining source was a small license operation in Hungary.

As well, arrangements had to be made for the timely selection and training of technical and flying personnel for the Fl 282, which was the responsibility of the Luftwaffe, as a certain core of personnel was required for service trials. (24)

A memo from the OKM/1.Skl. dated May 12, 1942 referred to early suitability helicopter shipboard trials, with special emphasis on operations from anti-submarine vessels; the helicopter was seen as an interim means of controlling the pursuit of submarines from the air. Training of personnel was to take place within the shipboard squadrons (Bordfliegerstaffeln). Placed in charge of the service trials was a Luftwaffe officer who possessed the required experience in shipboard operations under difficult conditions: Overleutnant Claus von Winterfeldt: more about him later. (25)

Flettner's limited production capacity (2

aircraft per month) made series production impossible in 1942, and in July the navy's request was raised to 110 aircraft (60 for anti-submarine vessels and 50 for other surface ships). Consequently the search was begun for another firm to build the aircraft – without success as it was to turn out.

The request for 110 helicopters did not include the Fl 282 U. Taking into consideration only those losses associated with housing the aircraft and the lack of servicing on board the submarine, the yearly requirement was placed at between 100 and 120. Great value was placed on the acquisition of this type (Fl 282 U). (26)

Flettner began experimental flights with a shortened version of the type (the Fl 282 V9, without tail boom and empennage) in July 1942, but the naval staff then dropped its request for the type. So much development work remained to be done that there was no hope of obtaining the type for the anti-submarine role in the foreseeable future. A simpler alternative solution presented itself in the form of the Fa 330 Bachstelze (Wagtail). The latter required much less work and accordingly its potential for production in quantity was much greater. (27)

Pilot Fuisting (?) receives the congratulations of the head of the Italian military delegation following his demonstration flight.

Flettner (2nd from right) during the demonstration flight.

In the course of the conference held by the naval inspectorate on September 8, 1942 it was suggested that, for simplicity's sake, the "Flettner Device Fl 282" be called Kolibri and that the GL/C-2 I officially introduce this name. (28)

Service Trials on the Drache

The diary of the Luftwaffe General Attached to the Commander in Chief of the Navy makes reference to trials with the Fl 282 aboard the ship Drache (Dragon).

The first deck-landing experiments in co-operation with Flettner began in August 1942 using the E-Stelle Travemünde's air traffic safety ship Greif. Conducted in Lübeck Bay, the trials with the Fl 282 V6 were a success. During the course of the trials the Drache reached speeds of up to 14 knots (approx. 28 kph). The helicopter made its fourteenth take-off from the vessel on September 9, 1942.

At the urging of the OKM in the course of discussions with the GL/C-E 2 I, it was decided to conduct the experiments as service trials as soon as possible. On September 12, 1942 the Naval Staff U III proposed conducting the trials from the minelayer Drache in the Mediterranean. As an important area of operations, the Mediterranean offered the promise of worthwhile missions for the Kolibri, and the tactical trials could be conducted in the more favorable weather conditions that prevailed there. As well, a submarine (U-617) was to be made available to the trials for two to three days.

The test missions which the OKM selected for the Fl 282 placed great emphasis

on the anti-submarine role and were formulated as follows:

(1) Determination of the most suitable camouflage scheme for the helicopter for conditions in the Mediterranean Sea. It is not known whether the camouflage requirements for aircraft that are to fly at low altitudes are the same as for aircraft operating at high altitudes. Low visibility is of great importance to the tactical success of the helicopter against the submarine.

(2) Determination of the most favorable tactic for approaching a surfaced submarine once sighted. The helicopter must strive to get as close to the submarine as possible without being seen or heard. In addition to approaching from the most favorable direction as far as lighting is concerned, experiments must be carried out in order to determine to which altitude a helicopter must go on sighting a submarine at a greater distance in order to travel to the immediate vicinity of the submarine in gliding flight.

(3) Determination of the ability to spot a submerged submarine in various sea conditions at various depths from different heights. As well, determine the maximum depth at which a submerged submarine remains visible in a calm sea under favorable lighting.

(4) Determination of range of contact between the helicopter and the anti-submarine vessel using the FuG 19 radio system.

(5) With what offensive stores can the helicopter be equipped in order to force a surfaced submarine to submerge.
Tactically, this requirement would arise when a submarine sailing on the surface is faster than the anti-submarine vessels operating with the helicopter. Only by the helicopter forcing the submarine to submerge – with the consequent reduction in its speed – will it be possible to drive the submarine into the grasp of the anti-submarine vessels.

(6) With what offensive stores must the helicopter be equipped in order to engage a submarine visible beneath the water.

Fl 282 V6, GF+YF, during trials on the minelayer Drache in the Mediterranean.

(7) Visibility tests from a submerged submarine against a helicopter in the air in various weather conditions.

The Fl 282 V6 and V10 (reserve aircraft) arrived In Trieste on October 1942. The Drache had been in dry dock there since May 7, 1942 undergoing repairs and modifications. In charge of the newly-formed Erprobungs- und Lehrkommando 20 (Trials and Instruction Detachment) was the previously-mentioned Claus von Winterfeldt, since promoted to Hauptmann. A Feldwebel pilot, Flettner company pilot Hans Fuisting (Leutnant of the Reserve) and three men from the technical personnel went on board with him.

Mining operations together with the ship Bulgaria began in the central Aegean in mid-November 1942 and lasted until the end of January 1943. Approval for this trial operation by the two Kolibris in the waters north of Crete was given by the Naval Inspectorate following the absence of English aircraft in the area for some months. However, on January 3, 1943 both ships were repeatedly attacked by enemy bombers and torpedo aircraft; no damage or casualties were sustained. The "Admiral Aegean," who was in command of the vessels, was subsequently requested "to permit the operation of the two valuable Fl 282s only when there is no prospect of enemy countermeasures."

Two methods of approach were tested in order to determine the optimal space requirement for landing on the platform:

(1) Straight-in approach into the wind with the ship under way, also when stopped.

(2) Perpendicular approach from the sea with the ship at ninety degrees to the wind.

The results of the service trails on the Drache were summed up as follows at a conference held by the Naval Inspectorate on February 10, 1943: "The Fl 282 can now be operated successfully as a shipboard aircraft for patrol and anti-submarine tasks, even from small ships."

On account of the poor state of deliveries – at the beginning of 1943 there were only 12 prototypes on hand, of which 6 had been idled by the company – the OKM lowered its target for the end of 1944 from 110 to 60 Fl 282s. The outlook for the navy providing skilled workers or production capacity to support Luftwaffe production was about zero. Representatives of the GL/C spoke of their efforts "to build and deliver the planned pre-production series of 90 Fl 282s" as quickly as possible. They also reported that negotiations were under way with other companies in an effort to find license manufacturers.

Further trials involving the Fl 282 were planned for early March 1943. To be conducted

After its Mediterranean trials on the minelayer Drache, the V6 carried out further service trials in the Baltic from the anti-submarine vessel KUJ 13 (April-May 1943). (Chapman Collection)

in the Baltic from the anti-submarine vessel KUJ 13 (Type KUJ 41), they were to further investigate the type's potential for anti-submarine use in areas beyond the Mediterranean. The trials began on April 24 in conjunction with the 21st U-Boat Flotilla (training flotilla) as part of the Kriegsmarine's ongoing "exercises for sea missions" in the area of Danzig Bay and the north tip of Gotland Island and lasted until May 15, 1943. The fleet exercises involved underwater attacks on simulated convoys, consisting of target vessels and escorts.

Once, while over a convoy, Hauptmann von Winterfeldt was able to make out all twelve submarines that had been sent against it. Von Winterfeldt lost his life on May 10, 1943, when he was forced to put the Fl 282 down in the sea in a position 35 nautical miles northeast of Christiansöya Island after running out of fuel. He was posthumously promoted to Major for his work in helicopter development. (30)

Bordfliegerstaffel 3/196 for Special Shipboard Aircraft

In order to accomplish the planned military use of the Fl 282, a front-line unit with the associated ground and air organization was necessary. For this reason on March 13, 1943 the Genst.Gen.Qu.2 issued the formation order for Bordfliegerstaffel 196; the unit was to be based at Kiel-Holtenau. Assigned to carry out the order was Erprobungs- und Lehrkommando 20 under Hptm. Claus von Winterfeldt. The order to establish this test and instruction detachment, to be based at Travemünde, was issued on October 1, 1942. Von Winterfeldt reported his task complete on October 14. The primary mission of the detachment was to conduct service trials for all special shipboard aircraft (Fa 330, Fl 282).

Not only was 3/196 responsible for all tactical and technical training and familiarization of aircrew with shipboard operations in the Fa 330 Bachstelze, but as well with ensuring the provision of materials and personnel for the ships and submarines for which the Fa 330 was planned. The Staffel was also responsible for continuing the trials with the Fl 282 begun by Erprobungs- und Lehrkommando 20, in order to bring this helicopter to full operational readiness. This involved retraining pilots possessing the basic pilot certificate and instructing the technical personnel on the Fl 282. The training given the mechanics placed special emphasis on the rotor transmission and mounting the rotor blades. It was estimated that conversion training would take approximately three months. (31)

The Genst.Gen.Qu.2.Abt. was informed that the Staffel had been budgeted on June 10. 1943. Staffelkapitän was Hauptmann Johannes Polzin. Completion of formation and operational readiness of the entire Bordfliegergruppe 196 was to the Luftwaffe General on the Staff of the Commander in Chief of the Navy on September 1, 1943. The Gruppe consisted of:

Gruppenstab	Aalborg
1/196	Aalborg (Ar 196)
2/196	Orbetello, Italy (Ar 196)
3/196	Kiel-Holtenau (Fa 330, Fl 282)

Author's note: Bordfliegerstaffeln 1/196 and 5/196 existed at the beginning of the war; however, due to the absence of a unified command there was no connection between the two units and consequently operations and supply were not centrally controlled. In May 1941 the F.d.L. proposed creating a Bordfliegergruppe for the two Staffeln to overcome this shortcoming. In September 1941 the OKM replied that it did not consider such a reorganization worthwhile at that point in time. It wasn't until March 17, 1943 that the order was issued for the formation of Bordfliegergruppe 196. 2/196 (responsible for auxiliary vessels and convoy escorts) was to be formed from the former 5/196. As mentioned above, formation of the Gruppe was complete on September 1, 1943: 3/196 was organized as follows:

(a) Main unit in Kiel-Holtenau.
(b) Sub-unit for "glider training" in Kiel Holtenau with 2 Kranich (Crane) and 2 Grunau-Baby gliders and two Fw 44 tow aircraft. Gliding activity began on October 1,1943.

(c) Sub-unit "Fa 330" (Kiel-Holtenau) for training navy and air force personnel to fly the Fa

330. This sub-unit also carried out experiments and training with the Fa 330 in the Chalais-Meudon wind tunnel in Paris under the direction of Oberleutnant Lothar Zöllner and shipboard training on the MS Elsass at Pillau-Marshafen.

(d) Sub-unit "Fl 282," initially at the Flettner factory at Berlin-Adlershof airfield, where pilots were trained by company pilot Hans Fuisting.

Sub unit 3/196

(Extract from the war diary of Bordflieger-Gruppe 196.)

September 1943

The training of pilots proceeded very slowly this month because there was only one Fl 282 on hand. Moreover there were differences with flight instructor Fuisting, who has command of the sub-unit there. This resulted in the handing over of the detachment to Ltn. Gottfried Wagner. Consultation with the naval inspectorate failed to achieve an acceleration of conversion training. Therefore, while refitting the MS Elsass the Staffel also installed a landing platform so that shipboard training might begin immediately following the conclusion of land training in Adlershof and the transfer to Kiel-Holtenau.

This was never to happen, however. Two groups with a total of 34 pilots and 12 technical personnel were trained on the Fa 330 Bachstelze on the MS Elsass in the periods from September 2 to November 15, 1943 and from February 1 to March 21, 1944. Apart from several periods spent in the Schichau Dockyard in Königsberg undergoing repairs, the ship was based at Pillau-Marshafen for the duration of the training.

October 1943

Fl 282 flight training resumed in Kiel-Holtenau with altitude and training flights, landing exercises on the landing platform (on land) and cross-country flights. As the Staffel had only three Fl 282s at that time, on October 4 the sub-unit transferred to Schweidnitz, where Flettner had relocated its operation. More Fl 282s were available for training and practice activities there. The flying program included local training, cross-country flights, stops at other airfields and night flights (on moonlit nights). The technical personnel were taken along as well, and were put to work in the Flettner factory to help with final assembly and speed up delivery of the Fl 282s.

Flights were carried out in wind speeds between 70 and 80 kph; the helicopter's high center of gravity resulted in major speed fluctuations in heavy gusts. The following entry appears in the war diary on November 19: "All pilots cleared for land-based operations. The month of November saw the highest total of flying hours to date. Final assembly of Staffel aircraft carried out by the sub-detachment itself. Three aircraft were completed. Planned work for December: acceptance flights and familiarization after transfer to Kiel-Holtenau. Powered and glider flights by the pilots."

December 1943

Installing the rotor blades on GF+YF aboard the anti-submarine vessel KUJ 13. (Chapman Collection)

The helicopter is still attached to the tether rope as it makes a "jump takeoff."

The transfer took place at the end of November. Further practice flights were conducted in the marine weather conditions prevailing in Holtenau. On December 15 there were five Fl 282s on strength. According to the so-called "operational readiness reports," which had to be submitted every ten days, this was the highest number of aircraft reached prior to the disbandment of 3/196 in July 1944.

In the course of the flights in December it was found that at outside air temperatures of less than 0 degrees Celsius, the oil in the engine and the upper and lower transmissions did not reach the required temperature. As a result, these became so cold that the engine began to vibrate severely and revolutions fell off markedly. According to the "Summary of the Accomplishments and Losses of Bordfliegergruppe 196," testing of the Fl 282 was not completed in November.

January 1944

Flight operations continued as the oil temperature problem was investigated, however these were seriously complicated by the poor weather situation (gusts to 95 kph, heavy fog, ceilings as low as 80 meters) and suffered lengthy interruptions. Attempts to raise oil temperature were suspended on January 19 after Flettner advised that there was a danger of overheating the cylinders and spark plugs.

February 1944

Further training and practice by the pilots concentrated on bad weather and cross-

Flettner pilot Fuisting giving instructions to a Luftwaffe pilot at Schweidnitz airfield (Silesia) in the summer of 1944. In the background interested school children look on.

country flights in the Fl 282 as well as deck takeoffs (jump takeoffs) and landings on the landing platform and bomb dropping (blind). Moreover, flights in gliders and powered aircraft (Bf 109, Ar 199, Fw 44) allowed the pilots to retain their proficiency. Plans to test the FuG 19 (ultra-short-wave radio, also used by the Fa 330) that month could not be carried out. The Staffel received a Vemag 2 vertical magazine for 5-kg anti-submarine bombs; according to E-Stelle Travemünde records, this was installed in the Fl 282 V20. Improvements were made to the mechanical bomb release after tests revealed its unwieldiness.

On February 5 the Gruppe submitted a proposal to the Luftwaffe General attached to the Commander in Chief of the Navy that 3/196 be moved to Pillau. Holtenau was frequently a target of enemy air attacks at that time, and it was only a matter of time before the Kolibris there were destroyed. Pillau was seen as a good choice as the Elsass, which had been earmarked for shipboard operations, was based there.

March 1944

More local training and cross-country flights (some of which were instrument flights) and practice bomb-dropping took place. As the Fl 282 had no bombsight, pilots at first had to refer to marks placed on the forward cockpit structure for various altitudes and speeds. The experimental installation of a cross-hair sight proved unsatisfactory, moreover the helicopter's high vibration levels presented some difficulties in acquiring the target.

The following information is taken from the report on the second conversion course: training was conducted on thirteen days (out of 49) in the period February 1 to March 21, 1944, in the course of which a total of 113 takeoffs and landings were made with a total flight time of 44 hours, 21 minutes. Considerable delays resulted from the need for frequent repair and overhaul of the Fl 282s.

On March 8, 1944 the F.d.L.(Ia) ordered the transfer of the Staffel to Pillau, which lasted from March 22 to 30. There was no flying whatsoever in this period.

Major Gerrit Wiegmink, the Gruppenkommandeur, was killed on March 24 while en route to the Commander of Aviation Denmark. The W 34 in which he was flying (T3+CB, W.Nr. 1436) was shot down over Sonder-Travers by a Mosquito. All four occupants of the aircraft were killed. Wiegmink's place was taken by Major Schmidt on April 24.

April 1944

All helicopter training that month was land-based, as the MS Elsass had returned to mine-laying duties in the Baltic on March 26. Flight operations consisted mainly of test-flying the Fl 282 V14, V18 and V20 for the purpose of extending the temporary flight clearances of these aircraft. A simulated marine rescue (picking up a man from a life raft and depositing him on land) achieved satisfactory results. The Fl 282 demonstrated its ability to fly with the additional load of 2 men.

The Naval Inspectorate called for the disbandment of 3/196 as early as March 19, 1944. Following the termination of production of the Fl 282 (and the Fa 223) in November 1943 and the Fa 330 in March 1944, in April the activities of the Staffel were restricted to the equipping of the remaining Type IX D 2 submarines with the Fa 330 Bachstelze and some limited training on the Fl 282. Operation of a full Staffel for this limited role could not be justified, and it was subsequently carried out by a "Kommando 330", an appendage of 1/196. (32, 32a)

May 1944

The war diary contains no entries for the month concerning flight operations by 3/196, although the operational readiness reports for the 10th, 20th and 30th of May each show four Fl 282s.

On May 2, 1944 the new Gruppenkommandeur, Major Schmidt, presented the Luftwaffe General attached to the Commander in Chief of the Navy with a proposal for the reorganization of the 196th Gruppe and the reallocation of duties for the Staffeln. 1/196 would take over the training unit's responsibility for providing shipboard detachments for front-line ships in the Baltic area of operations and as a special mission would undertake limited convoy escort and anti-submarine duties along the Baltic coast from its base at Pillau.

The Inglorious End of the Kolibris

It was proposed that 2/196 would provide the shipboard detachment for the Tirpitz (Norway). General training of new pilots as per the previous training program would take place at Aalborg. Special limited mission: coastal reconnaissance over the North Sea and Skagerrak as well as providing escort to B.S.O. convoys.

3/196 had to be disbanded as a Staffel formation as its roles were essentially experimental in nature and no new relevant mission could be found. Its remaining duties could be performed by a sub-detachment.

After some discussion between the Naval Staff, the Luftwaffe Operations Staff (Ia) and the Luftwaffe General attached to the Commander in Chief of the Navy, on May 27 the Gruppe received a positive response to its proposal for reorganization in the form of a telex from the OKL Operations Staff Ia Air (Robinson) Secret Command Matter no 11151/44. The reorganization began on June 1.

Below is the operational readiness report from May 11, 1944 which indicates the Fl 282 prototypes on hand with 3/196 at that time:

June 1944

This month saw the disbandment of the Staffel as part of the reorganization authorized by the Luftwaffe Operations Staff and the process was virtually complete by June 17. Consequently Bordfliegerstaffel 3/196 for special shipboard aircraft had ceased to exist. Some of the remaining Fl 282s were assigned to the E-Stelle Travemünde (for example the V 18), the rest were returned to Flettner.

The sub-detachment "Fa 330 Bachstelze" still existed; it was integrated into 1/196 and continued trials on board the destroyer Theodor Riedel until June 5. There were still plans for a final operational trial on the Scheer, but this was canceled by the Naval Staff on August 7. Thus the last element of 3/196, the Bachstelze Detachment, was disbanded and the remaining Fa 330s were delivered to the air park in Holtenau.

Fate of the Fl 282 after the War

There is no confirmed information concerning the fates of those Kolibris that survived the war relatively intact. According to statements made to the C.I.C. by Anton Flettner in the camp at Kransberg (Taunus) on June 29, 1945, two Fl 282s were flown from Schweidnitz to Bad Tölz in March 1945. One, flown by company pilot Fuisting, was hidden in a barn near Tölz, the other in a mill near Penzburg (approx. 15 km. west of Tölz), Following the arrival of the Americans, Flettner reported their hiding places to the military administration in Tölz. The Fl 282s were then taken to Munich and

Condition of Kolibris with Staffel 1/196:

W.Nr.	Base	Hours remaining before transmission housing renewal at 50 flying hours	Remarks
V10	Pillau4tb	about 2	rotor blades with manufacturer, flying with rotor blades of the V14
V11	Flettner	about 30	engine unserviceable
V14	Pillau	about 35	without rotors, see V10
V18	Pillau	about 40	serviceable
V20*	Pillau	about 25	serviceable

Fl 282 (V20?), machine captured by the English, fuselage framework with upper rotor transmission and clipped rotor blades on an English truck. The placard reads: FLETTNER KOLIBRI GERMAN HELICOPTER 1944.

Stuttgart, where they were demonstrated to Allied officers by German pilots. There is no mention of their subsequent transport to England or America or of the third Fl 282, whose frame was still in England in 1966.

According to an article in "Archiv", a publication of "Gruppe 66" which was founded in England in early 1966, the following Fl 282s survived (as of 1966): Fl 282 V 23 FE-4613 (A-019), also T2-4613, USAF Museum Dayton, Ohio; Fl 282 V 15 FE-4614 (A-011); Fl 282 W.Nr. 28368 College of Aeronautics, Cranfield, fuselage frame only.

The Americans initially coded their captured aircraft with an FE (foreign equipment) prefix and a three- or four-digit registration number. Following the renaming of the responsible department, after September 1945 the FE prefix was changed to T2.

It should be mentioned that the Werknummer 28368 quoted above cannot be correct; it is most likely an (incomplete) part number. The designations A-011 and A-019 also do not match the German prototype designations assigned the Fl 282. They are in fact the transmission numbers; for example the V 12 had number 03 and the V 20 number 04.

Subsequent to the publication of "Fl 282 Type Documentation" in Flugzeug Magazine, I received the color photos which Mr. Frank Marshall took for me of the remains of the English Fl 282 in the Midland Air Museum (Coventry). One of these shows the number "0.4" on the cover panel of the upper rotor transmission, which is kept inside after having been removed from the "fuselage torso," which sadly remains in the open. This "0.4" also appeared on the cover panel of the Fl 282 V 20, as verified by a black-and-white photo taken at the end of the war.

One can therefore conclude that the machine captured by the English, which is referred to as Werknummer 28368 in the relevant English aviation literature, is in fact the Fl 282 V 20 CJ+SN W.Nr. 282000020.

Center:

A Fl 282 captured by the Americans. According to English sources this is the V15 (FE-4614). Rotor transmission number 011.

Bottom:

Fl 282 V23, CI+TW, a two-seater, with rotor blades removed, US registration number FE-4613. V-shaped horizontal stabilizer with the last two letters of the aircraft's code visible beneath the stabilizer (the German nationality marking between the two letters has been overpainted).